Goat

Animal

Series editor: Jonathan Burt

Goat

Joy Hinson

REAKTION BOOKS

To Peter, with love

Published by
REAKTION BOOKS LTD
33 Great Sutton Street
London EC1V ODX, UK
www.reaktionbooks.co.uk

First published 2015
Copyright © Joy Hinson 2015

Printed and bound in China by 1010 Printing International Ltd

A catalogue record for this book is available from the British Library

British Library Cataloguing in Publication Data

ISBN 978 1 78023 338 3

Contents

Introduction

If you are short of trouble, get a goat.
Finnish saying

When I was about five years old I was taken to London Zoo, in Regent's Park, by my mother as a treat. Naturally we visited the children's corner, where among the rabbits and guinea pigs there was a small enclosure containing goats. It was possible then, as it is today, to enter the enclosure and touch the goats. I remember clearly that while my mother was distracted by one small goat, another approached me and started to nibble at my dress. I remember being fascinated: what kind of an animal eats clothes? My mother was horrified and tried to shoo the goat away. It paused for a moment, then ran at her legs, head down, and butted her with all the force it could muster. The goat cannot have been more than 45 cm (18 in.) high. I was totally captivated by this small, feisty animal trying to pick a fight with a creature so very much larger than itself.

Goats were once a common sight both in the countryside and on city streets. In the early twentieth century in New York City they could be found pulling carriages taking people on rides through Central Park. Goat-milk sellers walked the streets of Paris and other European cities with their goats on leashes, and the animals could be milked to order. Mollie Panter-Downes in her gloriously evocative novel *One Fine Day*, written in the immediate aftermath of the Second World War, tells us that 'goats were hopefully tethered on once-spruce lawns' in English villages.[1]

Tim Hunkin's Tug-of-war Goat Arch at the entrance to the children's corner of London Zoo, Regent's Park.

Goats, often so dismissively called 'the poor man's cow', have proved useful to mankind throughout the history of civilization. They provide not only meat, milk and hair, but also a means of transport. In developing countries they are used for haulage (they are surprisingly strong for their size), their dung is used as a fertilizer and their skins are used for carrying water. Goatskin also yields a particularly fine leather, called kid, which is perhaps best known for its use in glove making, giving us the expression 'to handle with kid gloves', meaning to handle very carefully. Morocco leather, also produced from goatskin, is used for book bindings. Parchment was a writing material obtained from untanned goatskin. It was historically used for many formal documents, including by universities for their degree certificates.

Indeed, many universities still refer to their degree certificates as parchments, although they are now no longer printed on goatskin. Untanned goatskin is also used to make musical instruments, particularly drums.

In some cultures goats are a form of currency, while in others they are used as wedding gifts. In the Hindu Kush the Kalasha people measure wealth and social status by the size of a family's goat herd. The larger the herd, the more likely a man is to be able to hold a feast and so gain added status within his community.[2]

Goats have been employed in some very imaginative ways – for example, in Roman times they were used to collect the fragrant resin of the shrub *Cistus ladaniferus*. The resin, called ladanum, is still used today in the perfume industry. According to Pliny the Elder, as the goats browsed on the shrubs their beards and shaggy legs became impregnated with the waxy resin, which could be combed out when they returned to their stables.[3]

Goat carriages in Central Park, New York, c. 1904.

In the developed world goats have been subject to the caprice of fashion. For long periods of history they have taken second place to sheep and cows, which historically yielded higher-value meat, milk and skins. Goats and their products have been useful during periods of privation, but have been cast aside in favour of the more valuable animals in more prosperous times. Perhaps only now, in the twenty-first century, is the value of goats being properly recognized. Over the past 30 years goat numbers around the world have grown at a tremendous rate, far surpassing the increase in the human population. In 1998 it was estimated that there were 700 million goats across the globe: a 52 per cent increase from the 1979 estimate. During this time the global human population increased by 33 per cent.[4] The increase in goat numbers

Paris street goats,
c. 1900–1910.

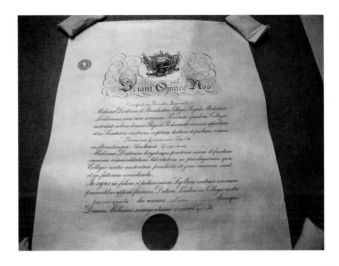

was greatest in Oceania and Asia, but there was still a 33 per cent increase in European goat numbers. By 2007 global goat numbers had increased to 850 million, approximately twice the number recorded only 28 years earlier.[5]

Goats have a symbolic significance: in early pagan cultures they represented lust and debauchery; in satanic cults they often represent Satan himself, while in Christian culture they symbolize sinners, those who have fallen from grace. Goats appear in the form of satyrs in Poussin's paintings of scenes from classical mythology, and as a representation of Satan in Goya's *Witches' Sabbath*. In other artwork goats can be found in the classic seventeenth-century pastoral scenes of Paulus Potter and in prehistoric rock paintings. Representations of goats as sacrificial animals appear in ancient Egyptian temple paintings and on Roman coins. In ancient civilizations both jewellery and everyday objects such as combs and razors were made in the form of a goat, and they

Bota de vino, a traditional Spanish goatskin wine bottle.

can be occasionally found in Japanese netsuke. More recently, nineteenth-century ceramics show figurines of goatherds along-side the more conventionally decorative shepherdesses.

The goat motif also appeared in twentieth-century art. Pablo Picasso cast a life-size bronze of a pregnant goat in 1950. At the time his mistress had just given birth to their second child,

Paloma, and Picasso was exploring classical themes. His she-goat figure is totally dominated by the distended abdomen and the udders formed from ceramic jugs. The goat is emaciated, with her neck barely supporting her head, but her eyes are half-closed and she appears content.

In heraldry goats appear in the coats of arms of several families, not surprisingly including the Kidd family arms, but also those of the Baker family. They feature prominently in the coats of arms of towns and districts throughout Europe. This usually reflects the economic significance of the goat in a town or district: thus the district of Aurland in Norway has a goat's

Bank of Israel wild goat coin.

Japanese netsuke.

Pablo Picasso, *The She Goat*, 1950.

Coat of arms of the municipality of Sachseln in Switzerland.

The coat of arms of the Haberdashers, one of the City of London's ancient livery companies. The Haberdashers made gloves, purses and other personal items.

head emblem because the district is known for its production of goat cheese, while the city of Bradford in England features an angora goat because of the importance of this animal in the local wool trade. Many towns in Switzerland have a goat in the town emblem, and the coat of arms of the Canton of Grisons is based on an ibex, a wild goat found in the region. Goats can also be found in the coat of arms of two City of London Livery Companies, the Cordwainers and the Haberdashers. The two 'goats of India' were added to the Haberdashers' coat of arms in 1570, reflecting the importance of goat products, particularly leather, in the haberdashery trade.

The word goat dates back to the Old English word *gát*. As a feminine noun this referred only to female goats, while males were called *bucca* or *gátbucca*.[6] From the end of the fourteenth century until the nineteenth century, the different sexes were generally called he-goat and she-goat, although the word buck specifically referred to a male goat until the sixteenth century. In the mid-nineteenth century, however, the term billy, a diminutive of the name William, began to be used for a male goat, and later that century the diminutive of Anne, nanny, was applied to a female goat.[7] Billy-goat and nanny-goat are commonly used today to distinguish between male and female goats, although breeders and biologists generally prefer buck and doe. The young of goats are called kids, a word from Middle English first used in about 1200. It was only much later, in the seventeenth century, that this word was also applied to children.[8] Castrated males are called wethers.

The language we get from goats is rich and varied, revealing our ambivalence to the goat. A scapegoat is the fall guy: a person who takes the blame for the failings or misdeeds of others. The adjective goatish is used to refer to general goat-like qualities, particularly the odour of goats. It is also used more specifically

to mean lascivious or lustful, for example in *King Lear*, where one character is described as a 'whoremaster man', who has a 'goatish disposition'.[9] In a similar vein the word nanny, as in nanny-goat rather than a person who cares for children, is used in the English vernacular to refer to a prostitute, with the old term for a brothel being a nanny-house.[10] By contrast, in Russian *koza* (nanny-goat) is an old-fashioned term for a tomboy. In several languages the word for a goat is used as a personal insult. In Germany *die dumme Ziege* is used in the same way as 'stupid bitch' is used to insult women, while in Russia *kozyol* (billy-goat) is a generally insulting term. In the Spanish-speaking parts of South America, *cabro* (billy-goat) is used as a homosexual insult in the same way as 'poof' and 'queer' are in Britain. In most Spanish-speaking countries *cabron* (goaty) is used as a moderate pejorative, and *cabrito* (young goat) is used similarly.

On the other hand, the phrase 'to play the goat', or 'play the giddy goat', means to play the fool or act irresponsibly, but it is

Goat clock at Poznań Town Hall, Poland.

used affectionately and is certainly not a pejorative. In Spanish the term *como una cabra* ('like a goat') refers to somebody who is crazy: in English we would say 'mad as a March hare'.

In both English and Italian there is the term capricious (Italian *capriccioso*), derived from the Latin word for a male goat, *caper*; it refers to someone who acts on a whim and whose decisions are subject to sudden change. The Latin also gives us the verb to caper, which means to leap around like a goat, historically used in reference to dance: 'to cut a caper'. Nowadays caper is more frequently used as a noun, which in its broadest sense means any kind of activity or escapade, but which also more specifically refers to dubious or downright illegal activities. A caper has become almost synonymous with a scam. Similarly, in Spanish *cabronada* ('like a goat') means dirty trick.

In Italian, the phrase *capra e cavoli* ('goat and cabbages') is the equivalent of the English phrase 'to have your cake and eat it', while in French the expression *ménager la chèvre et le chou* ('to look after the goat and the cabbage') means to sit on the fence or to have a foot in both camps. Goats' wool is referred to in the Italian phrase *sono questioni di lana caprini* ('they are questions about goats' wool'), meaning issues that are not worth debating. There are French equivalents of the expression 'to get one's goat', meaning to irritate: *prendre la chèvre* ('to take the goat') means to get angry, while *faire devenir chevre* ('to make a goat') is to enrage somebody. Very similar phrases exist in Spanish. Wine that is very acidic is said to be *vin a faire danser les chèvres* ('wine that makes goats dance').

Russia is a country with a long tradition of goat-keeping, so goats feature in several common expressions: *pustit kozla v ogorod* ('to let the billy-goat into the garden') means to let all hell break loose, or to set someone a wholly inappropriate task. *Ot nego kak ot kozla moloka* ('you will get as much from him as you would milk from a billy goat') refers to somebody who is useless. Curiously, in Germany the disease we call mumps is known as *Ziegerpeter* (goat Peter).

In medieval England 'goat' was used in a number of place names, through the Old English name 'gat-'. The names, which have lasted, are often evocative of the goat's ability to reach normally inaccessible places, so we have Gatescarth, which means goats' pass, in hilly Westmorland, and Gaterigg, or goats' ridge, in North Yorkshire.[11]

It is helpful to distinguish between wild, domesticated and feral goats. Wild goats are animals of the genus *Capra*, which are distributed around Afghanistan, the Middle East, North Africa and Europe. Domesticated goats are found all over the globe, on every continent except Antarctica. They all derive from wild goats, but were originally caught and kept by humans many centuries ago. There is a large number of breeds of domestic goat with a wide range of different appearances: some are smooth-haired, some woolly; they range in colour from jet black to creamy white; some have wide, flared horns, others bear very small, neat horns. This variety of goat breeds is the result of selective breeding. However,

all domestic goats remain one species, much in the way that the domestic dog has a wide range of physical forms, all of which belong to the same species. Feral goats are those that have been domesticated, then returned to the wild, either by escaping from captivity or by deliberate release into the wild.

Most domesticated animals feature strongly in popular culture, but goats are found only very rarely in cartoons, television programmes and advertising. In modern children's culture goats are sometimes portrayed as greedy animals, eager to eat everything within their reach. In many other children's stories they are shown to be rather naive and foolish. Let us take a closer look at goats and our relationship with them.

1 A Natural History of the Goat

The high hills are a refuge for the wild goats.
Psalm 104:18

We are much more familiar with domestic goats than we are with their wild cousins. How many people could name a wild goat species? Ibex are probably the best known of all the wild goats – they are certainly far better known than their cousins the markhor, bezoar and turs. Chamois are related to wild goats, but only very distantly. Sheep are much closer relatives. The problem is that so few of us have actually seen a wild goat, even on television. They are simply not as telegenic as the African elephant, the Bengal tiger or the giant panda. They also live in some of the most inaccessible parts of the world. However, they are impressive beasts with huge, threatening horns, and some really interesting habits.

Wild goats inhabit places in which few other animals can survive, climbing almost sheer rock faces in a way that seems to defy gravity, and leaping across boulders with apparent fearlessness. Their spectacular horns immediately distinguish them, and it is difficult to reconcile these handsome wild goats with their domesticated counterparts. Goats are ruminants, part of that broad family of hoofed mammals which includes giraffes, deer, antelope and cattle. More specifically, goats are members of the tribe called Caprini, and are most closely related to sheep and the Himalayan tahr.[1] Wild goats usually live at high altitude and at high latitudes in cold, mountainous regions, although the Nubian ibex is an exception, living in the hot deserts of the Middle East and North Africa.

Goats are thought to have evolved in one place and spread outwards from there. It is likely to have been a lowland location, possibly around the Mediterranean. The earliest sites of ancestral goat fossil remains are at Pikermi and Samos in Greece, and at Maragheh in Iran. The fossils found at Pikermi and Maragheh

Nubian ibex.

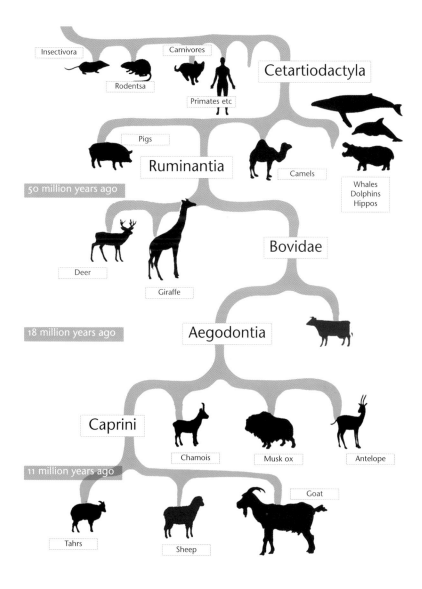

EVOLUTION OF THE GOAT

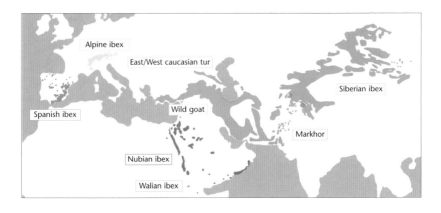

date to about 8.2 million years ago, while the remains at Samos are 7.2 million years old. The ancestral goat, *Pachytragus crassicornis*, was a little smaller than present-day wild goats, probably similar in size to domesticated animals.[2] *Pachytragus* had long, fairly straight horns, which spread slightly across the shoulders.

As goats were pushed further into areas of high altitude, evolutionary pressures would have favoured a larger body size, so wild goats became larger animals. Their horns became bigger, as the males with larger horns were most likely to win battles involving the clashing of horns with other males before mating. Their skulls also grew thicker, partly to be able to carry larger horns, but also to withstand the effects of clashing horns with other goats. The earliest true goat fossils date from 1.77 million years ago and have been found in Dmanisi, Georgia, interestingly co-located with remains of a species of early man, *Homo georgicus*.

There are nine recognized species of wild goat and most of these are types of ibex.[3] The others are the markhor (*Capra falconeri*), the East Caucasian tur (*C. cylicorndris*), the West Caucasian tur (*C. caucasica*) and the Persian wild goat (*C. hircus*, or *C. aegagrus*), which is the most likely ancestor of the domesticated goat and probably the earliest species of wild goat. These wild goats are distributed across Europe and Asia, with some being found in North Africa. There is little geographical overlap between the different species. Wild goats are medium-sized animals. The males typically have a body length of around 140–60 cm (55–62 in.), and stand at 80–110 cm (31–43 in.) at the shoulder. The males usually weigh 80–125 kg (176–275 lb). The markhor is the largest of the goats, with a length of up to 185 cm (73 in.) and a height of 115 cm (45 in.). The Spanish ibex is the smallest species, standing at only 75 cm (30 in.) at the shoulder, and the Nubian ibex is the most slender, with the males usually not exceeding 70 kg (154 lb). There are marked differences between the sexes, with

Siberian ibex.

the females being 20–30 per cent smaller than the males, although in the Spanish ibex the female is very much smaller than the male, at about half the weight.[4]

Wild goats have a woolly coat, usually consisting of two layers: an outer layer of coarse guard hairs and a thick under-layer of softer hair. This coat grows thicker in winter and is shed during the spring. There is often a change in colour of the coat between winter and summer. Mature male goats all have noticeable beards. All wild goats have prominent horns that are usually much longer in the male than in the female. Unlike deer, goats do not shed their horns each year, although the horns do show an annual growth cycle and a goat can be aged by counting the distinctive rings or ridges on its horns. Studies on the Nubian ibex have shown that the rate of horn growth varies during the life of a goat: in the first four years the horns grow at a rate of 12–20 cm (5–8 in.) each year.[5] After the fourth year the rate of horn growth slows, and after the tenth year growth is only 2–4 cm ($^3/_4$–1$^1/_2$ in.) each year. Horn growth is affected by several factors, such as disease, drought and nutrition. Stress is also a factor: a male ibex caught at the age of nine years had shown a steady horn-growth rate of 6 cm (2 in.) for each of the preceding years, but during the first year in captivity this was reduced to 2 cm ($^3/_4$ in.).

The life expectancy of wild goats varies between about ten and fourteen years, but it has been reported that the Caucasian tur and Siberian ibex can live for up to 22 years in captivity.[6] Wild goat species have a number of natural predators, including the leopard, snow leopard, wolf, lynx and eagles. They have also been extensively hunted by man throughout history.

When it comes to feeding, goats are browsing rather than grazing animals, sampling a wide range of vegetation including grasses, plant roots and thorny shrubs such as acacia. They are also superb climbers. Goat hooves have a large number of

Alpine ibex.

rough ridges on the underside and are very well adapted to gripping rocky surfaces. The horny outer part of the hooves grows continuously and is usually worn down as the animal walks over rough or rocky surfaces. In domestic animals kept on grassland the hooves need to be trimmed regularly. The hooves also function as shock absorbers, enabling goats to make their characteristic leaps from boulder to boulder. It is mainly due to the specific structure of the hooves that goats are able to feed on plants growing in inaccessible places. As a result they are able to live in places that do not support other grazing or browsing animals. Although wild goats are found in a wide range of habitats, from snowy mountainsides to rocky ridges and hot, arid deserts, the common feature of these places is a lack of competitive pressure from grazing animals.

Goats are mostly diurnal animals, active during the hours of daylight. However, in common with other ruminants they alternate periods of activity with periods of rest during the day. This time of rest is necessary to enable them to chew the cud, a process essential for the digestion of their rough herbivorous diet. Wild goats typically graze during the early morning and late afternoon, resting in the middle of the day, especially when the temperatures increase in summer. They usually rest on rocky outcrops, which provide a good vantage point and are relatively safe from predators, and move to areas of grass and scrub to feed. Goats usually live at a higher altitude than other grazing or browsing animals (often above the tree-line), and migrate to lower slopes in winter to avoid heavy snow.

Goats are social animals and generally live in herds that may be either single sex or mixed. The size and composition of these groups varies enormously, according to species, season and local environmental conditions.[7] Typical group sizes are two to ten individuals, although stable herds of up to 150 animals may

Goats on a nearly vertical dam wall.

occur. For most of the year males and females live in separate groups, which come together during the mating season, or rut. Prior to the rut the bucks fight among themselves – usually while still in their bachelor herds – to establish an order of seniority. The fighting involves the bucks wrestling with their horns and trading blows of increasing ferocity. Two well-matched bucks will stand facing each other, rearing up on their hind legs and

Male goats during rutting season.

crashing their horns together as their forequarters come down. The horns are very important in these displays. Studies have shown that males born without horns are unable to breed because they cannot compete physically for access to the females in a herd. There is therefore a strong selective advantage to having horns. The fighting usually decreases during the actual rut, with the breeding males avoiding each other.

Both male and female goats often show aggressive behaviour towards other members of the herd. This usually involves low-level butting and pushing, generally by socially superior members of the herd towards smaller, and younger, subordinates. Dominance is reinforced by the use of scent in male goats. It has been suggested that the raised tail typically seen in goats is significant, because the main scent-producing gland is in the tail. Goats use urine in a most striking way: while urinating, male goats adopt a hunched position with the penis extended towards the face or even in the mouth. As they urinate the face, beard and entire front quarters become drenched with the urine, producing a strong scent. It is likely that the scent of the urine conveys information about a goat's physical condition to other goats, both male and female.

Most species of goat have a clear breeding season, which ensures that the birth of the young coincides with the most favourable time of year for feeding. The exception is the Ethiopian ibex, the only member of the goat family to inhabit a hot climate. There is no particular advantage for the Ethiopian ibex kids to be born at a specific time of year, so this species breeds throughout the year. Domesticated goats reach sexual maturity within less than a year and have mostly lost the seasonality of breeding characteristic of wild goat species. Even while lactating domestic goats become sexually receptive and can thus have two litters each year, frequently producing twins or even triplets.

Wild goats reach sexual maturity when they are about eighteen months to two years old. In general the females reach maturity before the males. This contrasts with the maturation of domestic breeds of goat, in which the female reaches reproductive maturity, coming into regular oestrus well within the first year, often at between five and seven months of age. Wild goats in captivity become sexually mature at a much younger age than their wild counterparts. Domesticated males also reach puberty much earlier, and there are accounts of male kids impregnating females within three months of birth.[8] Although wild goat males are physically capable of reproducing by the age of two years, most only start to do so much later, when they are fully grown and in peak condition and thus able to compete effectively for access to females.

Female goats in the wild come into oestrus, the period of sexual receptivity, once each year during the rut. However, they continue to participate in the rut, bearing kids every year, until they die: there is no phenomenon comparable to the menopause in goats. Males, on the other hand, have a shorter reproductive life as access to females is determined on a competitive basis, and they are only in peak condition to compete in the rut for very few years. Male goats do not usually have a harem of females (although the Nubian ibex is an exception to this), but generally follow a single female in oestrus. Like other animals that have a regular oestrous cycle, goats are only sexually receptive when they are in oestrus. Male goats follow females during the rutting season and perform a ritual of crouching and vocalizing in order to induce the females to urinate. A male tastes the urine to determine whether a female is sexually receptive or not.[9]

The gestation period of wild goat species varies between 135 and 180 days, depending on the species, but is typically 150 to 170 days, compared with the 149-day gestation of domestic goats.[10] In preparation for giving birth the doe finds an isolated place in rugged and inaccessible ground. Goats mostly have single kids or twins, and very rarely triplets, each weighing between 2 and 3.5 kg (4–8 lb) at birth.[11] Goat kids usually remain hidden for two to four days, with the mother returning frequently to suckle the kids. After this time the kids follow the mother until they join up with other kids in a nursery group. Female goats rarely defend their young against predators. Weaning is often a gradual process, but is usually completed by the start of the rut when all suckling ceases.

Male goats are known to exert a strong influence on the timing of oestrus in females, stimulating adult females to come into oestrus in preparation for mating. The use of male goats specifically for this purpose is known as buck teasing, and it is used to synchronize oestrus in a herd of goats so that the kids are all born at the same time. The interaction between the male and females is complex and involves most of the senses: sight, smell, sound and touch.[12] The most important factor is the male: he must be sexually active. Females in mixed wild goat herds

Goat statue, Sierra de Gredos, Spain.

therefore come into oestrus when the males rut and become sexually active. Male sexual activity is determined simply by day length, with shorter periods of daylight causing the males to start the rut. This effect is mediated by a hormone called melatonin, which is released during periods of darkness and inhibited by light. As day length shortens after the summer solstice, so

melatonin levels increase in the male, which stimulates sexual activity. Implants of melatonin have the same effect and can be used to induce sexual activity during the non-breeding season.

Most wild goat species are considered to be under threat of extinction. The main threats to them are hunting and habitat loss. Wild goats are hunted for both their meat and their horns, which are highly prized as hunting trophies. As human agricultural activity extends into previously uncultivated regions, so wild goat populations are pushed into gradually shrinking areas. Some wild goats, most notably the markhor, are found in war zones, with the additional threat of military activity on population numbers.

The International Union for the Conservation of Nature (IUCN) has classified the West Caucasian tur as Endangered and estimated a population decline of 50 per cent in 1983–2004, mostly due to hunting.[13] There were thought to be 5,000–6,000 animals in 2004, but numbers are still decreasing. The East Caucasian tur is classified as Near Threatened and estimates suggest that there are probably about 10,000 mature animals left.[14] The IUCN lists the markhor as Endangered and estimates that there are fewer than 2,500 mature animals remaining with a very high rate of population decline: 20 per cent over two generations, with a generation estimated at seven years.[15] Each one of the fragmented groups of markhor contains fewer than 250 adult animals, which makes each group very vulnerable. The IUCN classifies the Persian wild goat as Vulnerable and estimated a 30 per cent decline in numbers over the last three generations.[16]

In 1986 the population of Nubian ibex was estimated at 1,200 individuals and the IUCN assessment was that the species was Endangered. Captive-breeding programmes and hunting restrictions over much of the species' range have helped numbers, and in 2008 the IUCN downgraded the level of concern, reclassifying

Markhor, the official animal of Pakistan.

Copper alloy
she-goat
from Turkey,
6th century BC.

the species as Vulnerable although there are fewer than 2,500 mature individuals remaining.[17] The Critically Endangered listing given to the Ethiopian ibex in 1996 has been downgraded to Endangered as numbers appear to be increasing slowly. However, there are thought to be fewer than 500 animals remaining in the wild, with none currently being held in captive-breeding programmes.[18]

The Spanish and Siberian ibex appear abundant, and there is no concern for their future. Indeed, the Spanish ibex is so numerous in some areas that it is considered to be an agricultural pest. Likewise, the Alpine ibex is classified as a species of Least Concern by the IUCN because there are significant populations spread over a wide geographical range.[19] Historically this species has been hunted close to extinction, and at the start of the nineteenth century there were only about 200 individuals left, which were found in one small area of the Italian Alps in the Valle d'Aosta. A combination of natural migration and reintroduction programmes has resulted in a significant increase in both their range and numbers, and it has been estimated that there are currently more than 30,000 mature individuals, with numbers probably still increasing.[20]

One effect of the isolation of wild goat populations is the limited genetic diversity of individual groups. In some areas conservation work has focused on the reintroduction of herds of animals into an area. These have been established with small numbers of individuals, which are often geographically very isolated and thus unable to interbreed with other herds and maintain a good genetic diversity. There is also concern about possible interbreeding with domestic goats, particularly in areas where ibex populations are located close to domesticated animals.

Domesticated goats have been developed into a great variety of different strains or breeds. The Food and Agricultural

34

Organization of the United Nations keeps a database of domestic animal breeds, which includes more than 1,150 breeds of goat.[21] These can be broadly classified according to geographical origins, so Swiss or Alpine goat breeds such as the Saanen are generally dairy goats, South African goats such as the Boer are meat goats and cashmere goats are used for fibre production. There are dwarf varieties of goat, found widely in Africa and Asia, which are also used for meat.[22]

As early as 1848 Henry Stephen Holmes-Pegler described a wide variety of different goat breeds and cross-breeds, even though goat breeding was an amateur activity in England at the time.[23] The goats in England were not thought to be worth the same level of breeding development that was applied to cows, sheep and even pigs. Across the world we have been generally slow to recognize the benefits of proper breeding and record-keeping. In Australia, for example, up to the late 1980s only three breeds of goat were recognized: dairy, feral and angora.[24] The use of a

Angora goat, from George Shaw, *Museum Leverianum* (1872).

stud book and the development of specialized goat societies, many dedicated to a single breed, have together done a great deal to establish the characteristics of specific breeds and to maintain healthy breeding stocks.

Physically the main visible difference between wild and domesticated goats is in their horns: domesticated breeds have been selected to have much shorter horns than those of their wild relatives. There is archaeological evidence to suggest that the earliest domesticated goats had scimitar-shaped horns, which were similar to those of ibex and the ancestral bezoar, but without ridges.[25] However, goats with short horns or no horns at all are much easier to handle than goats with long horns, and there is less risk of injury to their keepers. Thus modern domesticated goats, derived from many generations of selective breeding, do not have the magnificent horns of their ancestors. The Romans were the first to try to develop a breed of hornless (polled) goats – as Pliny notes, 'In both sexes those that have no horns are considered the most valuable.'[26] However, the Romans found that undesirable qualities came with the lack of horns: hornless goats were often hermaphrodite and infertile.[27] Despite two millennia of goat breeding there is still no such thing as a hornless breed of goat, and some goats are born with horn buds while others are not.

It is widely recognized that there are fertility problems with regard to the offspring of hornless goats.[28] If hornless males are mated with hornless females, an apparent gender imbalance results in the offspring, with far more males than females being born. Is there a 'lethal gene' that causes the prenatal death of female kids in these polled goats? In fact, it has become apparent that a proportion of the goats that were thought to be male are hermaphrodite females. They appear male but have two x chromosomes, so are genetically female. These 'intersex' individuals really are partway between male and female: typically they do not have a scrotum or udder, and the genital opening lies in between the location in a normal male (distant from the anus) and a normal female (close to the anus). Usually only small testes are present, which are retained within the body cavity, but sometimes there is both an ovary and a testis, resulting in a true hermaphrodite.[29]

Why is it so difficult to breed a hornless goat? The problem is one of genetics, and the great scientific advances of the twentieth century have finally allowed us to understand this issue.[30] When the gender imbalance of the offspring of polled goats was reported, it was not clear whether there was a genetic problem that was lethal to female foetuses. It was eventually recognized that the very low number of female kids counted was due to many of the females having the appearance of males, so that they were counted with the males. They were in fact the hermaphrodite form.

The issue is to do with the arrangement of genes in a goat's chromosomes. Goats have 60 chromosomes, arranged as 30 pairs, and the gene that determines whether a goat has horns or not is found on chromosome 1. This gene has two forms, a hornless (P) and wild type (horned p). The hornless form is termed 'dominant', so if one of the pairs of chromosomes has the hornless gene, then the individual goat has no horns. Very close to the hornless gene, in the same region of chromosome 1, is another gene that is important in fertility in female goats. This gene also has two forms, a 'wild type' (p), which gives normal fertility, and a mutant, often called polled (P), gene. The mutant gene (P) is inherited in a 'recessive' way, which means that if one copy of the gene is present the individual is fertile, but if both copies are present the animal is infertile. In fact, female goats with two copies of the P gene are usually hermaphrodite or 'pseudo-male',

Jacob Jordaens, *Goat*, c. 1657.

with rather underdeveloped male genitals, despite being genetically female. So all goats with horns are fertile, but if a female goat has no horns it has either one or two copies of the fertility gene, and may or may not be fertile. This also means that a goat with horns carries no copies of the infertility gene, so cannot pass it on to its offspring. As you need two copies of the infertility gene to get an infertile goat, it makes sense to breed from goats that have horns, and for all stud bucks to have horns. Of course, it is possible to manually remove the horn buds of very young goats. This is typically done when goats are about five days old. The horn buds can be treated with chemicals applied directly to them to prevent horn growth, so goats that are genetically horned can be manually polled and thus made safer to work with.

Logically it would surely make sense to eradicate the hornless form of the gene and eliminate the problem of infertility. It is not quite that simple. Curiously, while having two copies of the hornless gene makes a female goat infertile, having just one copy makes her *more* fertile.[31] It is far more common for a goat to give birth to twins or triplets if she has a copy of the hornless gene. So the ideal is mating a male with horns with a hornless female: half the offspring will have horns, but they will all be fertile.

It is worth briefly mentioning the Rocky Mountain goat. This species (*Oreamnos americanus*) is found in the USA and Canada, throughout the Rocky Mountains. However, despite its name, it is not really a goat at all. There are no native wild goats in any part of the New World. The Rocky Mountain goat belongs to the

A Rocky Mountain goat (this is not a 'true' goat).

An illustration of a
goat from Conrad
Gessner, *Icones
Animalium* (1587).

same tribe of ruminants as true goats, the Caprinae, but forms a
genus all on its own that is quite distinct from the genus *Capra*,
the true goats. *Oreamnos* translates from the Greek as mountain
lamb, and it is easy to see how it got this name. It would be a very
large lamb, though, as a full-grown male stands at about 1 m (3 ft)
high at the shoulder and weighs up to 135 kg (298 lb), and is thus
significantly larger than a fully grown sheep. In appearance the
Rocky Mountain goat is much like a true goat: it has a beard, very
short horns and a creamy-white woolly coat. It is a hardy animal,
living at high altitude and able to tolerate temperatures below -
40 °C. It has earned a reputation for aggression, as females have
been known to stand up to predators such as cougars, wolves and
bears when they threaten their kids.[32]

How *do* you tell the sheep from the goats? In Europe this is
generally not difficult, with domesticated forms being very different in appearance, sheep being woolly and goats usually having
short coats with straight hair. In China little distinction is made

between sheep and goats: indeed, the Chinese name for a goat simply means mountain sheep, and in the Middle East and North Africa sheep and goats can look remarkably similar. There are, however, a few characteristics that set them apart from each other. Perhaps the most useful distinguishing feature is the fact that all goats normally hold their tails erect, while the tails of sheep hang down. Where they have horns, these are of different shapes: a goat's horns are usually slender, sweeping backwards over the shoulders, then turning outwards, while a sheep's horns are chunky, usually starting to curl downwards at the back of the head and often forming spirals. Then there are the behavioural features: sheep tend to remain in one place as long as there is grass or other good pasture available, while goats tend to roam, sampling and browsing on different vegetation including shrubs, bushes and tree bark. Finally, when lambs are born they get up very quickly and follow the ewe, while kids stay in one place, usually in a sheltered hollow in the ground, to which the doe returns at regular intervals.

We have seen that the wild goats are mostly elusive animals, inhabiting mountainsides and other inhospitable areas. They are different in both appearance and habit from their domesticated relatives. However, it is from the globally nearly ubiquitous domesticated goats that we derive so much folklore, cultural practice and commercial utility.

2 The Domesticated Goat

You go to a goat for wool.
16th-century English saying, meaning to embark on a
foolish enterprise

Goats are tremendously useful animals: they do not need rich pasture like cows and sheep do, and indeed do not do well in such conditions, but thrive on poor scrubland where the vegetation or terrain is too rough for other domesticated animals. They thus can increase the amount of land that is brought into economic use. They are frequently described as the 'poor man's cow' and it is not difficult to see why. Like cows, they produce milk, meat and skins that are used to make especially fine leather. Contrary to the old English saying, goats do produce wool, often of exceptionally high quality, which can be harvested and woven into cloth. A goat also requires a lot less space than a cow: you can raise half a dozen goats on the land required for one cow.[1] A goat therefore makes an excellent contribution to a subsistence farmer's income: it can clear land of scrub and weeds, provide meat and milk, and breed readily to produce kids that can be sold. In developing countries the possession of a family goat can be the one factor that holds malnutrition at bay, providing protein and energy as well as calcium and vitamins.

Mankind has had a long relationship with goats. They have featured in artworks since humans began to paint images of themselves and the world around them. There are, for example, representations of ibex found in the caves at Chauvet, in the Ardeche region of France, which date from 30,000 BC; they are the very oldest cave paintings known. The later paintings at Pech Merle in Quercy (25,000 BC), at Lascaux (17,000 BC) and at Niaux (11,500–10,000 BC) also feature ibex. In the caves at Altamira, in the Cantabria region of Spain, most of the pictures show bison, but there are goats scattered among the images. In China, the earliest images of goats have been found on pottery of the Yangshao culture dating from 5000 to 3000 BC. Goats can also been seen in the later petroglyphs, pictures painted on boulders or rocks, dating from 1500 to 500 BC. From ancient Egypt, too, there are very early images of goats on a range of artefacts and in paintings. In some cases they are shown as sacrificial animals, demonstrating their significance in the early stages of civilization.

An African family and their goats.

2 Goat design from Banpo pottery, China.

Storage jar decorated with mountain goats, Iran, 4th millennium BC.

Archaeological evidence suggests that the goat was one of the earliest animals to be domesticated. Remains of a domesticated form of goat, with scimitar-shaped horns, have been found at Jericho, in Jordan, and dated to around 7000 BC.[2] Remains of a similar age have been found in Iran. By the fourth millennium BC scimitar-horned goats were found throughout the region. Artefacts depicting domesticated goats have been found from Mesopotamia, Sumeria and Egypt dating to around 2500 BC onwards. Goats were first introduced to Northern Europe during the Neolithic period, around 5000–3000 BC, brought by invaders through Switzerland and into Germany. At around this time a subgroup of domesticated goats, with corkscrew horns, emerged in Mesopotamia. It has been suggested that these goats were descendants of the markhor, the wild screw-horned goat, but the horns twist in opposite directions – the left horn of the markhor twists anti-clockwise, while that of the domesticated goat twists clockwise.[3] Ancestors of these goats with almost horizontal, twisted horns can been seen in many feral goat populations today.

Although wild goats are now found in quite small areas within Asia, Europe and Africa, domesticated goats are found all over the world, from the Arctic Circle to the smallest Pacific island. Genetic analysis of goat populations around the world, together with a study of Neolithic goat bones, has shown that, from the earliest time of domestication, goats have been moved extensively around the world, and there is evidence that domesticated animals have been cross-bred back with wild goats many times since their original domestication.[4] It is clear that from prehistoric times trade in goats was very active, and that it has continued ever since with significant intercontinental movement of animals. This contrasts with other domesticated animals, which have generally been confined to a small number of populations that have each established a distinct genetic character.[5]

Goats are particularly valuable where people inhabit marginal land, especially arid or rocky regions. Goat husbandry has allowed people to live in regions that would otherwise be largely unpopulated. The animals provide meat, milk and wool from land too poor or too steep for cattle or sheep. In the Bible we get an understanding of the significance of domestic goats in the Eastern Mediterranean region. Their importance is illustrated by the use of a total of ten different words for goat in the Bible (seven Hebrew and three Greek). Their milk was drunk and used to make cheese; their skins were used to make bottles for water and wine; and their hair could be woven into cloth for tents and

Paleolithic spear with a carving of an ibex.

45

Bronze goat, possibly from Corinth, 5th century BC.

Whetstone with handle in the form of a crouching ibex, from Iran, c. 900–700 BC.

shirts, and used as padding for pillows. All this was in addition to their use as sacrificial animals, with the accompanying rituals, symbolism and mythology.[6] The assumption is that nomadic peoples took goats with them, spreading the domesticated goat wherever they went.

We know that in the days of the great exploratory expeditions around the world, goats were taken on board ships to provide milk during journeys. Unlike cows, goats tolerated sea travel very well, so were an ideal animal for the purpose. It is believed that Captain Cook took goats on his first voyage to Australia.[7] As well as providing a regular supply of milk, goats were a valuable source of fresh meat. On a long voyage groups of goats were often carried by ships to be set loose on isolated islands, where they bred and formed feral populations that could be used to restock ships on their return journeys. The value of island populations of goats was recognized early in the days of global seafaring. In the journals detailing his third voyage, Christopher Columbus describes visiting the Cape Verde Islands in order to stock up on goat meat during the journey. He also noted that a particular large variety of goat was kept on the island of Boa Vista in order to provide skins which were salted and shipped to Portugal.[8]

Although it is not clear whether goats were included in the cargo of the *Mayflower* in 1620, they were certainly introduced to the Americas by the Spanish conquistadors, and reached North America very soon after the first settlers. As the American pioneers moved west across the continent they took goats with them. Not only did the goats provide meat and milk, but they could be used to clear scrub from farmland in preparation for cultivation. From these early imports of goats into the Americas a distinct breed developed: the Spanish goat. Until the 1840s this was the only goat found in North America and was distributed across the Southern states.

Domestic goats,
from *Bilder Atlas*
(1860).

In 1849 a small herd of angora goats was imported to South Carolina – the first pure-bred goats to reach North America. However, it was not until 1904 at the World's Fair in St Louis, Missouri, that dairy goats were introduced to the USA. Two Schwartzwald Alpine does were displayed, and in the same year the first goat registry was established in the USA. Over the next 30 years the USA imported a range of dairy goats, including Anglo-Nubians, Toggenbergs and Saanen. Domestic stock was cross-bred with Swiss dairy goats to form the American goat. Pygmy goats arrived as zoo specimens in the 1950s, but it was not until the 1990s that the first pure-bred meat goats came to the USA, with the arrival of both the South African Boer goat and the New Zealand Kiko goat in 1993.[9]

As goats spread around the world a number of different types developed. In Europe the main type is short-coated with scimitar horns and small ears, and is mainly used for milk production. Lop-eared goats are found from North Africa, across the Middle East and into northern India, becoming generally more long-coated towards the eastern end of this distribution. Most of these goats also have scimitar-shaped horns. From the mountains of Tibet to China the goats are mostly long-haired and often have cork-screw horns. In both East and West Africa a pygmy form is the main type of goat, while in South Africa the main form is a stocky lop-eared goat, which is mostly used for meat.

It can be difficult to assess the agricultural significance of goats. This is mainly because they have never been highly regarded as livestock, so were not always counted in agricultural surveys. In England, for example, the landscape is a key factor: with an abundance of rich pastureland the cow and sheep are able to thrive. A cow produces far more milk than a goat, and a sheep produces more wool. Both are relatively easy to contain in fields, and both produce highly valued meat. In general the goat thrives where

Cashmir goat and 'Tebetan goat' from *Bilder atlas* (1860).

Roman goat figurine made of copper alloy found in the River Thames.

'Amos and the Goats', an illustration from a late 12th-century English bestiary.

cows and sheep do not – on relatively unproductive land that is not covered in lush grass, but in scrub plants and woody shrubs. Goats in England have therefore historically occupied a position somewhere between that of the 'proper' farm livestock and that of domestic pets, while in Scotland and the mountainous regions of Europe and Asia, they are more highly valued for both milk and wool.

In the thirteenth century Thomas Aquinas classified animals into three groups: wild animals were classified as beasts, all domestic animals were called cattle and animals that fell into neither of these categories were termed quadrupeds. Goats were quadrupeds and therefore fell into that no-man's land between wild and domesticated animals.[10] This classification, together with the biblical portrayal of goats as representing sin, and the medieval view of goats as lascivious, has surely combined to work as a force against goat-keeping.

It is thought that goats were originally introduced to Northern Europe, including Britain, in the Neolithic period at a time when Britain was part of the continental landmass. During the Roman occupation of Britain more goats were brought into the country from the Mediterranean region. However, across Europe there is evidence that sheep displaced goats in many regions. Excavations of remains from the fifth to the seventh centuries in Britain have shown sheep and goat bones in a ratio of approximately 40:1, although excavations from the eighth to the eleventh centuries report ratios as high as 10:1 in Herefordshire.[11] It is possible that these remains underestimate the numbers as goats are usually eaten very young, and the bones of young individuals do not survive as well as those of older animals. However, written records from sources such as the Domesday Book suggest similar ratios of goats to sheep.

The Domesday Book of 1086 was a census of England at the time, recording not only human inhabitants, but also, in some counties, the livestock. From this we know that there was a significant goat population in England in the eleventh century but, as today, there were far more sheep recorded than goats. Goat numbers varied, as might be expected, according to the type of pasture available. Therefore in flat and grassy Cambridgeshire there were more than 90 sheep for every goat; in Devon, with its hills, woods and moorland, there were only seven sheep for every goat. The overall ratio of sheep to goats was 11:6, with a

'A Goatherd Defending his Flock', from the 9th-century First (or 'Vivian') Bible of Charles the Bald.

Leo the Wise,
'The Birth of
the Antichrist',
illustrating an
Italian manuscript,
1577.

total of 24,684 goats recorded. Extrapolating this for the counties
not included in the livestock survey suggests a total goat population of around 80,000 animals. The Domesday survey was far
from comprehensive, however, and is likely to have significantly
underestimated goat numbers because it recorded only those

animals held on demesnes, the large estate farms associated with manor houses, and did not include those held by the tenants, who occupied most of the land.

Later information on goat-keeping in England is found in the records and accounts of individual manor houses and monasteries, but it is clear from these that goat-keeping declined through the Middle Ages, with only very small numbers of the animals appearing in records from the sixteenth and seventeenth centuries. Bone records from Kings Lynn in East Anglia show that only 3 to 8 per cent of animals used for food were goats, but they also suggest that people ate the goat meat rather than just using the dairy produce. We know that kid meat was particularly valued among the nobility, but it also seems that goats never really overcame the prejudice against the image of them prevalent in the Middle Ages – that of being lustful and wicked.

It appears that there was a campaign against goats among influential landowners, who attempted to remove the animal from the land under their control, generally by charging a much higher levy for goat grazing than for sheep.[12] There were several

Jan Baptist Weenix, *A Goat Lying Down,* c. 1645–60.

factors that made goats unpopular with landowners. Goats were thought to foul the land and make it unfit for other animals to graze. Deer apparently disliked their smell, so the grazing of goats was prohibited in many royal forests.[13] Goats also eat young saplings and thus prevent regeneration of woodland. It was quite common for land-lease documents to include a clause that permitted grazing of livestock 'except for goats'. Taken together, these all fuelled the campaign against goats. There is an apparent paradox here because, as we have seen, the meat of kids was very highly regarded indeed. It is likely that herds of goats kept for kid meat were maintained on separate holdings so that they did not come into contact with other livestock. We can deduce this from place names such as Gatwick, meaning goat farm. This name occurs more than once in England, and suggests that there were farms specifically designated for goat-raising.[14]

There were clearly regional differences in goat-keeping that persisted after the Domesday survey. In the grassy lowlands of England sheep and cattle have always been the main livestock, so the tax records from the thirteenth and fourteenth centuries

Goats from Revd
W. Bingley,
*Memoirs of British
Quadrupeds* (1809).

show no references to goats in the home counties of Buckinghamshire, Bedfordshire, Suffolk and Sussex. However, in hilly, wooded and mountainous regions goats provided a useful source of both income and food. The number of local by-laws passed in the fifteenth and early sixteenth centuries, generally banning goats, suggests that goats were still commonplace in some areas. The active discouragement of goat-keeping must have had an impact on numbers, because by the late sixteenth and seventeenth centuries references to goats in estate inventories had become rare, although some goats were kept in Cornwall and in the far north of England. During this whole period it is likely that most goats were kept by peasants either individually or in flocks as part of the subsistence economy, rather than by the major landowners. There are many accounts of peasants being fined for trespassing offences by their goats onto estate land, although in some areas landowners rented estate land to local peasants for grazing, with rates for each goat varying between a farthing and a penny per year.[15]

What happened between the seventeenth and twentieth centuries? It is a great pity that the British government did not collect statistics on goat holdings in this period, when numbers of other agricultural animals were diligently recorded. However, there were several factors at work against the goat during this time, not least of which were the Acts of Parliament passed between 1750 and 1860 by the government. These 'Inclosure Acts' removed commoners' grazing rights from common land and open fields across Britain. In total, 28,000 square kilometres or 7 million acres were removed from common use, amounting to 21 per cent of agricultural land. It is inconceivable that this systematic removal of grazing land was without effect on goat numbers: it certainly had an enormous impact on rural people no longer able to graze their animals on what had previously

been regarded as common land. Although it became very much more difficult for ordinary people to keep livestock, there was a 'resistance movement' among cottagers, who kept goats tethered on the remaining wayside land. The Inclosure Acts led to a great number of goats being turned loose to fend for themselves, contributing to a significant feral population in Britain.[16]

During this period of time there were huge advances in livestock breeding and the development and improvement of sheep, pig and cattle breeds. It was the era of the 'Great Improvers' of British livestock, started by Robert Bakewell, who developed distinctive breeds of cow and sheep on his Leicestershire estate in the 1760s. However, the goat was not included in these breeding programmes and was largely neglected during this time, as it has tended to be during periods of economic prosperity.[17] In France there was a similar prejudice against goats because they caused much damage to vineyards and forests. It was only in the area of Mont d'Or that goats were the main livestock kept, and even in the mid-nineteenth century it required several thousand goats to keep the city of Lyons supplied with cheese.[18]

British goats, which tended to be bred only with other local goats, developed regional characteristics, and up to the early twentieth century it was possible to distinguish between the English, Welsh and Irish goat. The English goat is described as:

Head neat and tapering with moderate beard, horns set far apart, rising slightly at first with an inclination to the rear, then branching outward; ears rather large. Body long and

Johannes van Noordt, *Landscape of Cows, Goats and a Dog with a Milkmaid Beyond* (1644).

square-shaped with a fairly close coat. Colour ranges from black to white, but more often fawn, with a dark line along the back and black on the legs.[19]

Sadly this English goat can no longer be found. It has been re-placed or cross-bred with imported dairy goats. The last pure English herd of goats died out in the 1950s.[20]

Goats have always had their champions. In the early eigh-teenth century Richard Bradley, who wrote *A Survey of the Ancient Husbandry and Gardening*, expressed his regret that 'goats are not more general' in England.[21] However, Thomas Bewick, writing in 1792, noted that:

> The goat is an animal easily sustained, and is chiefly therefore the property of those who inhabit wild and uncultivated regions, where it finds an ample supply of food from the spontaneous productions of nature, in situ-ations inaccessible to other creatures.[22]

London Zoo, 1831.

It is clear that the goat was considered an animal for 'wild and uncultivated' land, not a major part of domestic livestock in 'cultivated' regions. In the nineteenth century Henry Stephen Holmes-Pegler made a valiant attempt to raise the profile of the goat, in 1848 publishing *The Book of the Goat; Containing Full Particulars of the Various Breeds of Goats and their Profitable Management*.[23] Holmes-Pegler remarks that the 'naturally roving disposition and well-known mischievous propensity of the goat are its great drawbacks', and acknowledges that goats are at a disadvantage compared with sheep, which can easily be pastured in herds without constant attendance. He does suggest, however, that the grass by the roadsides in Britain may be 'turned to good account for pasturing goats belonging to cottagers living close by'. He accepted entirely that there will never be commercial goat farming in Britain because of the lack of demand for goat milk.

In the novel *Lark Rise to Candleford*, which reflects the author Flora Thompson's upbringing in the 1880s in a rural hamlet on the Oxfordshire border with Northamptonshire, a typical pastoral region of England, there are no references at all to goats,[24] suggesting that the regional pattern of goat-keeping apparent in medieval times had persisted to the Victorian era. By the time that the fourth edition of Holmes-Pegler's book was published in 1909 there had been some significant advances in British goat-keeping, including the establishment of the British Goat Society in 1879.[25]

During this time there were some curious developments. English country houses surrounded by parkland frequently kept herds of deer, but in some places in the nineteenth century these were either replaced by or combined with goats. These goats were neither truly wild nor in any way domesticated, but can perhaps be thought of as semi-feral. Perhaps the most famous park goat herd in Britain is the one that used to live in Windsor Great Park during the nineteenth and early part of the

Great Orme goats.

twentieth centuries. The herd was started by George IV, who accepted a gift of cashmere goats from the squire of Weald Hall, Essex, Mr Christopher Tower.[26] Mr Tower had imported a herd of goats from Kashmir in the early 1820s and in 1828 had their fibre spun into a shawl, which won the Gold Medal of the Society of Arts. The king was impressed, so Mr Tower made him a gift of a pair of goats, which were kept in Windsor Great Park. This pair bred and the numbers increased, but inbreeding was a problem. New breeding stock was introduced by exchanging goats with a herd kept by the Duke of Buckingham at Stowe Park, and later, in 1889, by a fresh herd imported from India. The Windsor Park goat herd survived until 1936, when it was removed to the Zoological Society of London's headquarters in Regent's Park, and later to Whipsnade Zoo in Bedfordshire, where it remains.

In the 1890s some of the Windsor Park goats were acquired by Major-General Sir Savage Mostyn, who bred them and released the resulting small herd on the Great Orme, just outside Llandudno in North Wales, where they still live. During the First World War, when battalions of the Royal Welch Fusiliers were being raised in Llandudno, an attempt was made to capture some individual goats so that they could serve as regimental mascots. The story goes that half the herd jumped over the cliffs to its death in order to evade capture, so the attempt at their seizure was abandoned.[27] Numbers of goats on Great Orme Head have fluctuated over the years. In 1953 there were only about 25 individuals, but by 1990 the numbers had grown and steps were taken to relocate 26 goats to Hereford and to the island of Flat Holm.[28] There was an outcry involving the local press and resident groups, and the local council reconsidered its approach and halted the control programme. By 2001 there were 250 goats in the herd and they were so short of food that they had become a

Bagot goats.

nuisance and were raiding local gardens. Popular opinion was much less opposed to the control measures by this time and the local authority took action, using a combination of contraception and rehoming. A total of 85 goats were relocated around England by 2007. Numbers of goats on Great Orme are presently controlled by a programme of contraception, carried out by both the RSPCA and the British Army, who use the herd as a source of regimental goats.

Probably the most ancient park herd in Britain is the Bagot goat herd, which dates back to the fourteenth century.[29] According to Bagot family tradition, the original goats were brought to the family home at Blithfield Hall, Staffordshire, between 1377 and 1380, as a gift from Richard II, who acquired them while returning through Europe from a crusade.[30] An alternative theory suggests that the goats were brought to England during the Norman Conquest. Recent genetic analysis of the goats proposes a third possible origin: that these goats originated in Portugal and were brought to England by John of Gaunt's army returning from the wars in Castile in the late fourteenth century.[31] The Bagot goat bears a striking similarity to the Schwarzhal breed from Switzerland, so it is also possible that it was brought from Switzerland.

Whatever its origins, the Bagot goat herd was established in the 32-hectare (800-acre) parkland around Blithfield Hall at least 600 years ago, and a goat has appeared on the Bagot family coat of arms since 1380.[32] It is likely that the original stock has been cross-bred with feral British goats over the years, although it has retained its very distinctive appearance. The Bagot is a fairly small goat, the males typically reaching no more than 70 cm (28 in.) at the shoulder, and is strikingly coloured with a jet-black head and neck, and a pure white body with long hair. Both males and females always have horns, which are slender, spread in a curve across the shoulders and very pointed. As is usual among goats,

Boy with a
Toggenberg goat,
1941.

the male has longer horns than the female. Temperamentally these goats are of a 'nervous disposition'. They have a significantly lower birth rate than most goats, as well as poor maternal skills, so their numbers have always been small.[33]

The herd at Blithfield Hall was gradually dispersed in the years following the death of Lord Bagot in 1962. Rounding up the animals proved challenging as they had spread well beyond the original parkland into surrounding woodland and country-side.[34] In 1979 there were only 61 goats remaining: 33 females and 28 males.[35] These were distributed by the Rare Breeds Survival Trust to a number of approved farms around the UK, although some have now been returned to Blithfield Hall.[36] Numbers are still very low – it is estimated that there are currently 200–300 breeding females, and the Rare Breeds Survival Trust classifies the breed as vulnerable.[37] One of the problems faced by the Bagot goat is that it has little commercial significance. It has spent most of its history as a park goat, where its decorative properties were all that mattered, but has no value as either a dairy or meat goat. Its existence remains precarious, and is continued by a small group of farmers who have a real affection for this rather splendid animal.

The relationship between Britain and India during the Victorian era had a significant impact on goat-keeping in England. The administrators of the British Empire became accustomed to goat milk while in India and began to keep goats on their return, in order that they could have a continued supply. This was facilitated by the custom of the ships of the East India Company of carrying goats on board as a source of milk. The original goats were replaced with fresh ones at ports en route to England. On arrival in England the goats were often housed with retired employees of the East India Company. In this way a number of novel breeds of goat were introduced to England and interbred with native goats. This created an interest in goat breeding and was one of the key drivers for the foundation of the Goat Society. The impact on the productivity of the English goat was astonishing. Over less than a century the introduction of 'a few dozen' goats into the 'mongrel hotch-potch' of the English goat herd led to the development of six distinct breeds and an increase in milk yield from an average of 250–300 litres (440–528 pints) a year to a maximum of around 1,800 litres (3,168 pints).[38]

There was a surge in goat-keeping during the Second World War as milk, cheese and other dairy produce was subject to rationing, but this declined rapidly after the war. The need to increase food production was a major concern for the post-war British government, but goats did not feature at all in its plans, even when considering the most difficult and unproductive land. A treatise published in 1953, *Marginal Land in Britain*, discussed ways of increasing land productivity in hilly or other marginal areas, but goats had fallen so far out of favour in Britain that they did not merit even a single mention.[39] Stud records from the British Goat Society show a slump in goat-keeping between 1945 and 1960. In 1945 there were just over 3,000 goats registered in the stud record. By 1960 this figure was 1,040. Numbers remained low throughout the 1960s, but by 1974 had increased back to post-war levels, and by 1978 there were 7,924 goats recorded.[40]

More recent figures show that goat-keeping in England continued to increase in the late twentieth and early twenty-first centuries. It is interesting to consider the factors that have driven this surge of interest. In the 1970s it was reported that goat-keeping in Britain was on the increase, but that many people with smallholdings tried to keep goats in the hope that they would act as 'a lawn mower and weed controller', rather than for their meat or milk, although an increasing demand for goat milk and other dairy products was also noted.[41] Around this time

Goat-milk street
seller, Palermo,
Sicily, c. 1906.

Goat-milk street seller, Palermo, Sicily, c. 1906.

there was a popular movement in Britain advocating a return to self-sufficiency, as illustrated by the BBC television comedy programme *The Good Life*, in which a very middle-class couple turn their suburban garden into a smallholding, complete with chickens and a pig, but no goat. This not only reflected a movement in post-1960s Britain, but stimulated greater interest in self-sufficiency, which was encouraged by the British government. With increasing concerns about factory farming and food additives in the 1980s and '90s, accompanied by a number of health scares including mad cow disease, even more people have turned to self-sufficiency in the late twentieth and early twenty-first centuries. A pair of goats has been seen as a key part of this lifestyle. Much easier to handle than a cow, and requiring less space, two goats provide ample dairy produce and good-quality manure.

The official government data show the scale of this rise in goat-keeping. Estimated figures for total goat numbers in England in 1970 were around 18,000, increasing to around 50,000 at the turn of the century and 86,000 in 2009.[42] While hobby farming accounts for part of this increase, there has also been a surge of interest in goat milk, partly prompted by the 'alternative' health lobby supported by celebrities such as Oprah Winfrey. Goat milk is now produced commercially on a scale that Holmes-Pegler would never have imagined, and is available not just in health-food shops, but in supermarkets across Britain. There are several large goat-milk producers in the UK. One commercial goat farm in Yorkshire, in northern England, keeps a herd of 3,500 goats.[43] The goats are milked twice daily in an automated milking parlour with 72 milking points. The farm has its own processing facility and supplies milk, cream, butter and yogurt to supermarkets

St Helen's Farm products.

across England. The behaviour of goats, with their roving and escapist tendencies, continues to present a challenge to farmers, so large-scale goat farms tend to keep their animals indoors in custom-built rearing sheds. It is therefore most unlikely that flocks of goats will become as familiar as cows and sheep on a country-side walk. Indeed, when travelling through the English countryside it is easy to get the impression that cows, sheep and horses are the only livestock in England.

At the end of the first decade of the twenty-first century, goat numbers in England are once more back up at the levels seen when the Domesday Book was compiled. There is an astonish-ing variety of goat cheese available from both supermarkets and specialist shops, and it is found on the menu of every smart restaurant. Goat numbers are thriving, with an increase in both large and small goat holdings. If Henry Stephen Holmes-Pegler was alive today he would surely be amazed at the rehabilitation of the goat, and satisfied that his vision for the goat in England is at last being realized.

While goat dairy products are valued very highly in Europe and other parts of the developed world, the same is not true for goat meat. As already mentioned, historically goat meat, and particularly the meat of young kids, was highly prized in Britain, with kid meat served as a delicacy at feasts and important cele-brations.[44] Bolton Priory in England kept a flock of goats and sent kids as gifts to the Archbishop of York in the early fourteenth century. Durham Priory similarly was buying kids to send to the households of the royal family and the archbishop. The meat of 50 kids was served at the coronation feast of King Henry III in 1221.[45] Clearly goat meat, or at least kid meat, was a valuable commodity, as it was noted in the sixteenth century: 'Yonge Kyddes flesshe is praysed above all other flesshe. Olde kydde is not praysed.'[46]

The French word for butcher, *boucherie*, is derived from *bouc*, which means male goat, reflecting that in France also goat meat was widely eaten, chiefly by poor people.[47] That is not the case in Europe today. It is not usually possible to find goat meat in the major supermarkets of Britain, France or the USA, although specialist producers supply an Internet-based mail-order niche market. In the most highly respected recipe books from Europe's greatest cuisines, of France, Italy and Spain, there is almost no mention of goat meat, and it is very rarely featured on the menus of smart restaurants, in marked contrast to goat cheese, which is ubiquitous.[48] In Britain there has been a great resurgence of interest in cooking, and a revival of some of the excellent old cooking traditions, particularly in relation to meat, championed by Hugh Fearnley-Whittingstall – yet even in his book goat meat does not merit a single mention.[49]

Only the Italian cookbook *The Silver Spoon* explains how to prepare *Cosciotto Pasquale*, Easter leg of kid, along with *Cosciotto alla Piemontese* and kid cutlets. Even so it notes that kid is 'difficult to obtain outside Italy' and suggests lamb as a substitute.[50]

The situation is very different in Africa and on the Indian subcontinent, where goat meat is a staple. Indeed, in Europe it is usually only in 'ethnic' butchers and restaurants that you can find any sign of goat meat, and even then it is seldom seen because goat meat is very rarely described as such. It is more often sold as *chevon*, which refers to kids with a carcass weight of around 15 kg (33 lb), and *cabrito* or *capretto*, which is young milk-fed kid of up to 12 kg (26 lb) weight. *Capretto* is a very fine and tender, pale-coloured meat that is only obtained from unweaned kids.[51] After weaning the meat changes colour, becoming both darker and less tender, and having a more pronounced flavour – it is rather like the difference between veal and beef. The more mature goat meat has been likened to 'bad venison',[52] but it is this that is used in Caribbean cooking. On the Asian subcontinent the term mutton can be used interchangeably for both sheep and goat meat.

Goat meat is not prohibited by any religious observance; quite the opposite, as goats were traditionally used as sacrificial animals in religious ceremonies, when meat was distributed to the people. In some cultures goat meat is still highly valued. While it is less often used as a day-to-day meat in the way that beef and lamb are, it can still be found at festival times, particularly in Muslim communities. In the Islamic tradition goat meat is eaten at Iftar, the meal with which the daily fast is broken during Ramadan. It is also a part of the festivities at Eid al Fitr, the festival that marks the end of the month of Ramadan. In the Christian culture goat meat is sometimes eaten at Easter, particularly in Greece and sometimes in Italy. In Caribbean culture goat curry is a staple dish.

Goats at
St Helen's farm,
East Yorkshire.

Zhao Mengfu,
Sheep and Goat,
c. 1300.

There is a significant global trade in goat meat. The world's largest exporter of goat meat is Australia, with an output of nearly 15,000 tonnes of meat and 138,000 live goats in 2001–2.[53] By 2005 the amount of meat exported had risen to over 20,000 tonnes. The largest importers of goat meat are the USA, the UAE, China, Qatar, Saudi Arabia and, perhaps surprisingly, France. In the USA and France this meat does not usually find its way to supermarkets, but is destined for the 'ethnic' market.[54]

Goat meat is generally tasty, although kid can be rather bland. It does not usually have a strong 'goaty' flavour unless older un-castrated males are eaten. It is low in fat and calories, so meets the needs of the health-conscious market. While goat meat is not easy to raise organically as goats are very prone to parasitic infections, goats do not respond well to intensive farming and are usually free range, so meet the requirements of ethical con-sumers. They also require less food and space than cattle, so are an excellent ecological and economical option. So why do West-ern societies consume so little goat meat? Food, like so many other things in Western society, is subject to fads and fashions. Perhaps, as the environmental and health benefits of goat meat are recognized, the fashion will change in favour of goat meat in the same way that it did first for goat cheese, then for goat milk.

The main value of goats across the globe is in the milk they produce. It has been estimated that more people in the world drink goat milk than cow milk. Globally, the largest producer of goat milk is India, where annual production of 4 million tonnes of milk was recorded in 2008, followed by Bangladesh with 2.17, Sudan with 1.5 and Pakistan with 0.7 million tonnes.[55] France, with its large goat-cheese industry, was the world's fifth largest producer, with 0.58 million tonnes. Although India is also the second largest global cow-milk producer, with 4.4 million tonnes, after the USA with 8.6 million tonnes, none of the other three major goat-milk producers has significant cow-milk production, and goat milk is certainly the major dairy product in these countries. China, the world's most populous country, is only a minor producer of goat milk: 0.26 million tonnes compared with 3.6 million tonnes of cow milk. Over the past ten years there has been a steady increase in goat-milk production, with a 30 per cent increase in consumption in all the major producers. Despite the enormous recent growth in interest and the widespread availability of goat milk in the UK and USA, neither of these countries features among the top twenty global goat-milk producers, although both are in the top ten cow-milk producers.[56]

Goats are astonishingly efficient at the process of converting rough forage into milk. It is not unusual for individuals from the major dairy breeds to produce more than 10 per cent of their body weight in milk every day. A prime Saanen goat can produce around 2,500 litres (5,283 pints) of milk per year (a typical lactation period is 305 days), working out to over 8 litres (17 pints) of milk per day, which is more than 12 per cent of her body weight.[57] Such records are, of course, achieved by goats from highly specialized breeds, maintained under perfect conditions. In France the average yield is around 400 litres (845 pints) per goat per year, which works out at about 1.3 litres (3 pints) per day through the lactation period.[58] St Helen's goat farm in East Yorkshire estimates that each goat yields 3 litres (6 pints) per day.[59] In developing countries milk yields are usually significantly lower than those obtained in Western Europe and the USA. Historically, in the days before specialized goat breeding, the average yield was only around 250 ml ($^1/_2$ pint) per day.[60]

Most of the milk produced in Europe is used in the cheese industry, which historically was concentrated in Switzerland and France. Over the past two decades goat-cheese production has increased in many other countries, notably in Britain and America. There has been a great increase in hobby farming in the USA, and at the start of the twenty-first century there were more than 250 artisan makers of goat cheese in the country.[61]

In the past it was considered that goat milk and cheese were too strong to be digestible.[62] In the late eighteenth century Thomas Bewick wrote of the milk:

From the shrubs and heath on which it feeds, the milk of the goat acquires a flavour and wildness of taste very different from that of either the sheep or the cow, and is highly pleasing to such as have accustomed themselves to its use.[63]

There remains today a current of opinion that goat milk is rather an acquired taste. This may have arisen from the view that it is smelly and 'goaty'. If milking does are kept in a herd well away from any male goats, and if the milk is cooled very rapidly after milking and used while fresh, then there really is no significant goaty flavour.[64] However, mature male goats really do smell very strong indeed, and if they are kept with females their scent invades everything, including the milk.

The other goat products that are particularly highly valued are cashmere and mohair, among the most expensive natural fibres available. They are prized for both their warmth and their softness. Mohair is produced specifically from the angora goat, a very distinctive breed with shaggy ringlets of hair. Confusingly, angora fibre does not come from the angora goat, but from a rabbit!

Cashmere, on the other hand, comes from a variety of breeds of goat, selected for high-grade fibre production, although there is a recognized cashmere breed. These goats, producing relatively large quantities of high-quality fibre, originated in the Kashmir region of India and Pakistan, and the name cashmere reflects these origins. There is a particularly fine version of cashmere called pashmina, produced by pashmina goats. Goat hair grows in a

Pashmina goats.

double layer. The outer layer consists of coarser guard hairs, while the undercoat is of much softer, finer hair: the cashmere fibre. In the winter this layer becomes much thicker and is usually shed in the spring. The hair is harvested either by clipping just before the animal would naturally shed its coat, or by combing as the coat is shed.[65]

The angora goat is probably the oldest of the currently recognized goat breeds, dating back to around 2400 BC (Yalcin). The name angora is derived from Ankara, a city in the Anatolian region of Turkey that was once the centre of the mohair industry. Mohair was so valuable to the Turkish economy that none of the angora goats were allowed to be exported until the nineteenth century. This goat breed was first reported by a Dutch visitor to Constantinople in the seventeenth century, who noted that the goats were kept as pets for the ladies.[66] In the nineteenth century exporters went in search of the fabled angora goat, and exportation of mohair fibre to Europe began in 1820. The establishment of mohair spinning in England in 1835 generated great demand for this luxury textile. In 1836 angora goats were exported to South Africa and in 1849 to the USA. These countries, together with Turkey, are the main centres of mohair production today. Angora goats are clipped annually in the springtime, much like sheep, but then clipped a second time in the autumn. The ringlets grow to 20–30 cm (8–12 in.) long between clippings. Castrated males are the best fibre producers, and typically an angora goat produces hair weighing up to 25 per cent of its body weight each year, an astonishingly efficient production rate of mohair.[67]

3 The Feral Goat

The goat is much more hardy than the sheep; and is,
in every respect, more suited to a life of liberty.
Thomas Bewick, 1790

Being natural escape artists and very adaptable creatures, goats
have not surprisingly formed feral populations around the globe.
The term feral refers to wild animals that are domesticated, and
are later either released deliberately or escape into the wild, where
they breed and establish self-sustaining populations. In the era of
great explorations from the fifteenth to the nineteenth centuries,
feral goat populations were established around the world. The list
of islands with feral goat populations is extensive and global, from
the Falkland Islands and islands around Vancouver, Canada, to
Mauritius and the islands around New Zealand.[1]

Due to the characteristics of the goats that were selected during the process of domestication – fecundity and adaptability,
combined with an innate hardiness – they very easily form feral
populations of large herds in areas to which they are introduced.
Goats have a quite remarkable breeding potential, and a feral
goat population can easily increase in number by 50 per cent
each year.[2]

Goats have been widely blamed for damage to the native vegetation in many places. Islands are particularly vulnerable, with
their restricted space and often delicately balanced ecosystems.
There are usually fewer predators to control goat numbers on
islands than on continental land masses, so the goat population
of an island is more likely to increase there unchecked. It is widely

The St Helena Ebony (*Trochetiopsis ebenus*) in bloom on the island of St Helena: only two plants, hidden on a sheer cliff face, survived the goats.

recognized that goats cause deforestation. However, it takes rather more than a few goats to turn woodland into bare scrub. Although goats do eat tree bark, they present little threat to mature trees. What goats are more responsible for is preventing the regeneration of areas where trees have been felled by human activity. They eat saplings, so these never have the chance to mature into trees. It is thus the combination of human and goat activity that is responsible for the problem, causing goats to come into conflict with people, especially with those whose livelihoods are threatened by goat damage to crops, and more recently with conservationists, who seek to preserve native species and habitats against goat damage.

Probably the earliest recognition of the problem that feral goats can cause to native species on an island was on the Atlantic island of St Helena.[3] Goats were introduced to this largely forested island by Portuguese sailors in the early years of the sixteenth century. By 1580 there had been a population explosion and there were descriptions of flocks nearly a mile long. The goats on St Helena all derived from a very limited gene pool, and in 1676 more goats were brought to the island to improve the strain. Of course, these introductions also contributed to the vigour of the feral population. The goats caused a great deal of damage to the vegetation, bringing about the extinction of some native plant species, such as the St Helena ebony, which was once abundant on the island.[4]

The first attempt to rid the island of goats was made in 1731. A ten-year eradication programme was started, which was repeated in 1745. This was partly successful, but as already noted just a few goats can breed rapidly and quickly repopulate an area. By the start of the nineteenth century the goat population was estimated at just under 3,000 individuals and another eradication programme was started. The forests on St Helena had largely disappeared by this time. Trees had been felled for timber, and goats had prevented saplings from maturing to replace the felled trees.[5] Further attempts to exterminate the goats were made in the mid-twentieth century, and these succeeded in reducing the numbers to low levels. St Helena is now officially free of goats and work is underway to propagate and reintroduce native plants in order to restore the island's vegetation.

One very simplistic view of nature conservancy is that all introduced species should be systematically eliminated to allow populations of native species to recover. So are feral goat populations always a bad thing, and what, if anything, should be done about them? The answer to that question rather depends on exactly where the feral goats are located and how long they have been there. In Crete, Australia and the Galápagos Islands there are significant populations of feral goats that are each being managed in very different ways.

The Cretan wild goat, which is also called the *kri-kri* and *agrimi*, was thought to be a unique subspecies of the wild goat, found only on Crete and some small islands off Crete in the Aegean Sea. It was therefore accorded highly protected status. This goat is a symbol of the island of Crete, featuring in tourist information and even on postage stamps. There are references to it in the Greek myths: it is the symbol of the vine cult that started on Crete and spread to ancient Greece. Archaic artefacts bearing the image of the goat have also been found on the island.

Recent genetic analysis has shown that the Cretan wild goat is, in fact, a feral goat descended from domesticated stock introduced to the region during the time of the Minoan civilization, which flourished between around 2700 and 1500 BC.[6] This feral goat is found in very limited parts of the Greek islands, but its main base is in the White Mountains of western Crete at the head of the Samaria Gorge. The gorge, located on the south-western coast of Crete, was classified as a UNESCO Biosphere Reserve in 1981, in recognition of the unique flora and fauna of the region, and specifically citing the wild goat.[7]

The IUCN survey of wild sheep and goats of 1997 reported that the Cretan wild goat (*Capra aegagrus cretica*) was 'vulnerable' to extinction, but that the conservation efforts on Crete 'appear to be working', with more than 600 individuals recorded. The same report recommends that 'feral Caprinae should be exterminated unless there are defensible reasons for making an exception.'[8]

There are clearly some defensible reasons in the case of the Cretan wild goat: this ancient example of a feral goat has been isolated from outside genetic influences for many centuries and thus has developed a unique character. However, it is described as being at risk from cross-breeding with modern breeds of

Cretan wild goat, or *kri-kri*.

Gold pendant in the form of a Cretan wild goat, from 1500 BC.

domesticated goat that are kept on Crete for their milk. The natural history of goats with the blurring of species boundaries and the ready cross-breeding between different 'species' suggests that this strain of goat deserves special consideration. Its iconic status on Crete raises it far above the status of just a 'feral goat'. Extermination or active protection? The choice is really not difficult here.

In Australia, where no native goats are found, early settlers in the 1780s brought goats with them to provide milk and meat. Escaped domestic goats formed feral populations. Some years later, in the mid-nineteenth century, cashmere goats were imported into Australia in an attempt to start an Australian goat-fibre industry. When the market collapsed in the 1920s, many of these animals were simply set loose and joined the feral goat population, reinvigorating the gene pool. In addition, in the twentieth century goats were employed to keep planted forests and grazing land free of weeds. Some escaped, some were abandoned and some were deliberately released. All these factors contributed to the establishment of an extremely large and well-established feral goat population, which has spread across all the states of Australia. It is so large that in 1996 it was estimated that there were 2.6 million feral goats in Australia.[9] To put this into some perspective, at the same time there were more than 100 million sheep in Australia.[10] Over the past few years parts of Australia have experienced severe drought conditions, and the goats have caused considerable damage to wide tracts of land: overgrazing combined with low rainfall has resulted in loss of vegetation followed by soil erosion. Competition with domestic livestock, particularly sheep, has cost the Australian sheep industry an estimated AUS$ 17 million. There is also a fear that feral goats may act as a vector in the spread of a number of potentially devastating livestock diseases, including foot and mouth, scrapie and blue tongue, if these diseases ever reach Australia.[11]

As a result, in Australia the feral goat has been designated a 'Class 2 animal' under the Land Protection (Pest and Stock Route Management) Act 2002, and all landowners are required by law to control goat numbers on their land.[12] However, despite an extensive eradication programme numbers have not substantially declined, and in 2010 it was estimated that there were still 2.3 million feral goats across the country. In Queensland there are an estimated 240,000 goats. A control programme in the state removes 150,000 animals each year. However, this programme has been totally dependent on the market value of live goats, their meat and hides, so is only fully operational at times when the market is buoyant. When the value of goats is low there is little culling and the numbers of goats increase. The rate of increase is quite staggering: unless actively culled, the population doubles every 1.6 years.[13] More recently, feral goat hunting has been promoted as a recreational sport in Australia and particularly in New Zealand, where a similar problem exists. The New Zealand government permits year-round goat shooting and offers online advice for hunters.[14]

The official programme of feral goat eradication from the most environmentally sensitive areas of Australia employs a variety of methods. In flat regions motorcycles are used to round up herds of goats, which are then taken to an abattoir. Goats are often found clustered around sources of water, particularly during periods of drought. Collecting yards have been constructed around watering holes, using goat-proof fencing and an ingenious system of one-way gates that allow the goats into the yard but not out again. In this way whole herds have been rounded up. These goats are either slaughtered or redomesticated – they make good breeding stock for both the Australian goat-meat industry and the mohair trade. The goat carcasses have a significant market value for both pet food and human consumption.

It is only recently that the value of the feral goats has been recognized as a source of breeding stock for refreshing the gene pool of commercial goat herds. Feral goats are generally more hardy than the more highly selected breeds of domestic goat, and able to withstand harsher conditions. They can thus be used to improve the hardiness of domestic goats. The relationship between the Australian authorities and their feral goats is therefore somewhat ambiguous: they have a clear economic value for both meat and hides, and also as a stock improver. However, when their numbers are uncontrolled they present a serious threat to both the environment and the established sheep industry.[15]

The situation in the Galápagos Islands is rather different from that in either Crete or Australia, with feral goats being a recent and highly destructive introduction. In the Galápagos an ambitious and extensive eradication programme has been undertaken by the Charles Darwin Foundation and the Galápagos National Park.[16]

The Galápagos Islands form an archipelago in the Pacific, off the coast of Equador. They were made famous by the writings of Charles Darwin, who visited the islands in the nineteenth century and made a series of observations on the wildlife found

Australian feral goats.

on each island. From these observations he developed his theory of evolution by natural selection. The wildlife of the Galápagos Islands therefore has a special place in the history of our scientific culture. Goats are not a native species anywhere on the Galápagos, and it is thought that they were introduced to the islands in the nineteenth century, probably by whaling fleets, and later replenished by local settlers.

Introduced goats cause a variety of problems that combine to threaten the native wildlife of these isolated and scientifically important islands. The appetite of goats is fairly indiscriminate for vegetable matter, and the goats significantly reduced vegetation cover on the islands, causing high levels of erosion. They also compete with native animals for food, and through these actions threatened to destroy the biodiversity of the islands.

Isabela Island is the largest of the Galápagos Islands and is home to more native species of plant and animal than any other island in the Galápagos. Many of these native species are seriously endangered. Isabela Island has distinct north and south sections, joined by the Perry Isthmus. Northern Isabela has a land area of around 250,000 square kilometres (97 square miles), and up until 1970 there were no goats at all here. It is not clear whether they were released in the area from shipping boats, or whether they crossed the Perry Isthmus, but in 1970 the first goats arrived. They bred very rapidly and by 1997 aerial and land surveys estimated a population of between 100,000 and 150,000 goats in northern Isabela. The whole of the Galápagos archipelago is a UNESCO World Heritage Site due to its biodiversity and scientific significance, and the goats in northern Isabela posed a problem that could not be ignored.

The Charles Darwin Foundation and the Galápagos National Park combined forces to address the problem. As a result, at the beginning of the twenty-first century a major project was started to eliminate the feral goat population of northern Isabela. The eradication methods were tested on the smaller islands of Pinta and Santiago between the end of 2001 and 2004.

The first stage of the eradication process involved shooting large groups of goats using helicopter teams. Once the numbers had been reduced, tracker dogs were used to round up small groups of goats ready for hunters using rifles with telescopic sights. Later stages of the programme, when goat numbers were low, used what is called the 'Judas goat' technique to find any remaining animals.[17] The Judas goat, named after the disciple who betrayed Jesus to the authorities, acts to 'betray' the location of small herds of goats. The term Judas goat was originally used for the goat that was employed at both shipyards and abattoirs to lead groups of cattle or sheep either on board a ship, or to slaughter. The Judas goat technique as used in conservation involves releasing onto the island a number of sterilized goats, each fitted with a collar that permits radio tracking. As goats are herd animals that naturally seek out other goats, these Judas goats find any remaining goats on the island and betray their location to the hunters. Using these techniques, Santiago and Pinta Islands were both declared goat free by 2005, fortunately before any of the native plants on Pinta had become extinct. Even for the relatively small island of Santiago, with an area of 586 square kilometres (22 square miles), this was a massive undertaking, involving two helicopters and more than 37,000 hours of fieldwork just to reach the stage where the Judas goats were introduced. This stage was only reached after 90 per cent of the goat population had been killed.

Northern Isabela was also declared goat free by the end of 2006, with just a few Judas goats remaining. It is now hoped that the natural habitat will regenerate and the native plant and animal species will be protected. Certainly, this has worked on other islands. Project Pinta was finally completed in 2010. Following the eradication of the goats the vegetation recovered quickly, and in May 2010 giant tortoises were reintroduced to Pinta after an absence of 40 years.

4 The Goat of Myth and Folklore

By candlelight even a goat looks beautiful.
French saying

Goats are woven into the traditional stories and belief systems of many cultures around the world, reflecting the global distribution of domesticated goats since early civilization. Goats feature in a wide range of traditional writings, from the Sanskrit Upanishads to the Welsh Mabinogion, but are very often incidental to the main narrative rather than the stars of the piece – in general they have no more of a role than as providers of transport for the main characters of the stories. It is interesting to note that in the traditions of the areas of Asia where wild goat species are found, there is very little evidence of goats in the mythology of the regions.[1] However, a wide range of stories from Europe feature goats in religious practice and folklore. The different stories present a varied picture of mankind's view of goats. Sometimes goats are purely utilitarian, as in the Norse myths, while in others they have distinct personalities, as in *Aesop's Fables*. In the Greek myths gods may take on the form of goats, and Satan is often represented with the head of a goat. Goats are seen to both represent evil and ward it off. In some stories the goat is wise or cunning, and in others it is quite the opposite. There is no single and consistent image of the goat.

Curiously, the goat is one of the animal forms most commonly represented in chimera – those creatures of myth and fable made up from various parts of different animals. In biology the

term chimera is used for an organism that contains a mixture of genetically different tissues, but the original chimera of classical mythology was a clearly defined creature, generally female, which included the body parts of a goat. Homer records that the creature, as also shown on a Hittite building at Carchemish in ancient Anatolia, had the head of a lion, the body of a goat and the tail of a serpent.[2] This was a representation of the sacred year: the lion symbolized spring, the goat summer and the serpent winter. Later the goat part of the chimera was lost and the sphinx emerged.

The most ancient chimera is the part-fish, part-goat deity called Capricorn, which dates back to the Middle Bronze Age, around 2000 BC.[3] The image of Capricorn began with the ancient

Sumerian god Enki, who became the later Babylonian god Ea. Enki was sometimes portrayed as a goat and sometimes as a fish. These symbols were combined to give the chimeric Capricorn. It has been suggested that Capricorn was originally a goat that jumped into the sea to escape Typhon, the monster of Greek mythology, and became half-fish.[4] Capricorn became the constellation Capricornus, depicted in Mesopotamian star catalogues.[3] The constellation of Capricornus can be found in the southern sky within the group of zodiacal constellations. In the old astrological calendar the sun was in Capricornus at the winter solstice, but in the modern calendar it passes through this constellation in late January and early February. The Tropic of Capricorn is the name still given to the line of latitude that marks the furthest south point of the sun at the winter solstice.[5] The brightest star in Capricornus is a double star, visible with the naked eye, called Algedi, meaning goat. The star Capella (Alpha Aurigae) is also known as the Goat Star, although the origin of this name in not clear.[6] Curiously, there is a group of annual meteor showers called

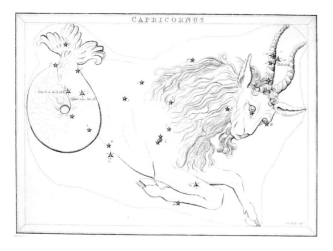

An image of Capricorn, c. 1800.

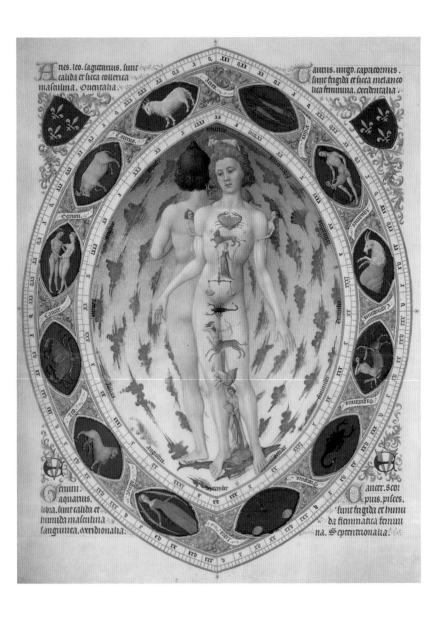

the capricornids, which appear to originate in the constellation of Capricornus and can been seen in three groups between April and September each year.

The symbol of Capricorn, depicted as either a goat-fish or a goat, is still used as the tenth sign of the zodiac. The Chinese zodiac also includes a goat, or sometimes a ram, as its eighth sign. People born in the year of the goat are supposedly intelligent, calm, creative and dependable. These are interesting character traits to associate with the goat, which is not often portrayed in such a positive light.

Corinthian bronze satyr, a human–goat chimera.

Of all the chimera we are probably most familiar with the faun, as illustrated by Mr Tumnus in *The Chronicles of Narnia*.[7] He has the lower body of a goat, complete with tail and cloven hooves, and the chest and head of a man, with the addition of small horns and a goatee beard. Mr Tumnus is a benign character who provides shelter and aid. The faun is a variation of the classical portrayal of the Greek god Pan and his attendant satyrs, who have a darker side than Mr Tumnus and are altogether more ambiguous.

Pan is the god of shepherds and their flocks, of the countryside and of peasant music. His father was Hermes and his mother may have been Amaltheia, the goat.[8] Like the fauns, Pan has the legs and tail of a goat, with a man's chest and arms, and a head with horns and a beard. He was said to be so ugly when he was born that his mother ran away and Hermes took him up to Olympus for the gods' amusement. He is often shown holding a set of pan pipes, the musical instrument with which he is most closely associated. Pan lived in rural Arcadia rather than on Olympus, and through his association with the rural environment he is also the god of fertility and hence associated with sexuality. Images of Pan often show him with a phallus, and he had a reputation for his sexual appetite. The myths that feature

The signs of the zodiac surround anatomical man, showing the human body in relation to the constellations, in the 15th-century French illustrated manuscript *Les Très Riches Heures du Duc de Berry*.

Pan playing
his pipes.

Pan include tales of how he, or his attendant satyrs, seduced or attempted to rape a large number of nymphs, minor goddesses and other females. He was also thought to be the agent that causes sudden fear in crowds, which is why he gave his name to the word panic. As punishment for his sexual misdemeanours Pan died, the only member of the Greek pantheon to do so. At the festival of Lupercalia, held in ancient Rome but persisting in pagan culture until the late fifth century, possibly in honour of Pan but perhaps related to an earlier fertility rite, goats were sacrificed. Their skins were cut into strips by the celebrants, called the Luperci, who adorned their naked bodies with the strips of goatskin, had a feast, then ran naked around the Palatine hill, hitting young women with the goatskin.[9]

The stories about Pan and his entourage form the basis for some of the fertility cults of Europe, and are part of the long association between goats and sexuality, and especially with male sexual depravity. Perhaps this association has resulted from direct observations of the behaviour of both wild and domesticated goats. According to Pliny goats are 'capable of generating in the seventh month and while they are still suckling'.[10] This early observation is borne out by more recent observations that male kids of domestic goats become sexually active at a very young age and will attempt to mount their mothers or any other female goat, with one report of a male kid impregnating his mother and sisters at the age of less than three months.[11] It has also been reported that the Nubian ibex, which is found around the southeast Mediterranean, becomes so sexually aroused during the rut that it masturbates by taking the tip of its penis into its mouth.[12] As William Blake wrote in *The Marriage of Heaven and Hell*, 'The lust of the goat is the bounty of God.'[13] Even the *Oxford English Dictionary* remarks that the goat 'is especially noted for its . . . lively and wanton nature'.[14]

In the thirteenth-century Chartres Cathedral, France, there is an allegorical sculpture depicting lust (Luxuria): a woman is tempted by the Devil, shown as part-goat, part-man. At Freiburg Cathedral, Germany, which also dates to the thirteenth century, the figure of lust (Voluptas) is shown as a naked woman draped only in a goatskin complete with horns and hooves.[15]

From the early seventeenth century the English language has used the word goat to denote a licentious man.[16] Botanically, the goat's reputation has been used to name a plant, a member of the *Epimedium* genus, family *Berberidaceae*, popularly called horny goat's weed. According to tradition, a farmer noticed that his goats became sexually active after eating the plant, although given the natural behaviour of goats it is tempting to wonder how he noticed. The result is the plant name and, with it, its reputation for aphrodisiac properties.[17] In China, plants of this genus

Pan and the she-goat, from the Villa of the Papyri, Herculaneum.

Hendrick van Balen the Elder, *Pan Pursuing Syrinx*, c. 1615–20.

have been used to treat impotence since the Han Dynasty (202 BC–AD 220). Finally, when considering sexual associations with goats, there is the most extraordinary saying from Afghanistan: 'A woman for duty, a boy for pleasure, but a goat for ecstasy.'[18]

In several cultural traditions goats appear as the embodiment of gods or mythical spirits. As well as stories about Pan, Greek mythology includes stories about the god Dionysus, who was often depicted as a goat. According to legend Dionysus was the illegitimate son of Zeus, and he was rescued by Hermes and transformed into a goat kid to escape from Hera, Zeus' jealous wife. He was raised in the form of a goat on Mount Nysa, where he invented wine.[19] Dionysus became the god of wine and his followers spawned a number of hedonistic cults. Offerings were made to him to protect the grape harvest, usually sacrifices of goats. Sir James George Fraser, author of the famous work on mythology and folklore, *The Golden Bough*, reports that there was a statue made of bronze in the form of a goat in one of the wine-growing regions of Greece. The local vineyard owners covered the statue with gold leaf as an offering to Dionysus, to protect their vines from blight.[20] Eventually Hera found Dionysus in his goat disguise and sent him mad, roaming the Earth accompanied by his tutor Silenus and a group of satyrs and maenads.[21] The maenads are a group of wild women, every one of whom Pan claimed to have seduced. Both Silenus and satyrs are strongly associated with goats: Silenus is shown dressed in a goatskin, and the satyrs are depicted with horns, tails and short, pointed goat ears, and have a reputation for being priapic and hedonistic.[22] These characters, together with the fauns, their Roman

Antony van Dyck, *Drunken Silenus Supported by Satyrs*, c. 1620.

counterparts, came to be associated with spirits of the woodland. It is not difficult to see why: goats, unlike sheep, have a taste for tree bark and other rough forage, so have a tendency to wander off into woodland, where, in dappled light, they might easily be thought to be woodland spirits.

This association between the form of the goat and woodland spirits is also seen in north European folklore, in which the Russian wood spirits are called Ljeschie, which means wood beings. The Ljeschie are also chimeric, taking the form of a half-goat and half-human. They are well known as shape-shifters, taking on themselves a tall form in woodland and a miniature form in the stubble of cornfields. This description of Ljeschie changing size to move easily between woodland and cornfield also illustrates the close association between goats and the mythology of corn and the harvest.[23]

The corn spirit is part of an ancient belief system across Northern Europe, which held that there was some primal life force within the corn crop that was killed by harvesting. The spirit of the corn took on an animal form, which in Northern Europe was very often the form of a goat, and certain harvest rituals were observed. The rituals reflected the local belief about the corn spirit, which varied from place to place. In parts of Bavaria the last sheaf of corn to be cut had two horns arranged on it and was called the horned goat, while in other parts of Germany a wooden goat was carved, decorated with oats and called the oats' goat. This effigy was taken home by the last reaper to finish harvesting his patch of oats.

In some regions it was believed that the corn goat lived in the farmhouse over the winter, so each farmhouse had its own corn goat. In other areas there was a single corn goat that moved from farm to farm as the corn was cut. In Skye, for example, the farmer who was first to finish harvesting sent a sheaf of corn to

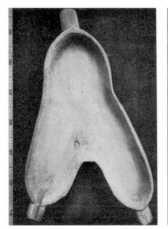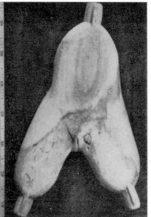

Wooden tray in the shape of a goat's udder.

his neighbour, who passed it on when his harvest was finished, and so on until the sheaf had completed the round of the local farms. The corn sheaf was called the cripple goat, possibly reflecting the idea that cutting the corn cripples the goat, the corn spirit.

Sometimes it was believed that the corn goat was killed by the reapers as they collected the harvest, so the harvest was accompanied by the killing of a goat. In the Alps, around Grenoble, France, a live goat was decorated with flowers and ribbons and allowed to run through the fields just before the end of the harvest. The reapers caught the goat, which was beheaded by the farmer as the most senior farm worker, then cooked and eaten as part of the harvest supper. However, a piece of the goat meat was taken and pickled for keeping until the next harvest, presumably acting to preserve the goat spirit until then. The skin of the corn goat was made into a cloak for the farmer, who wore it at harvest time if bad weather threatened. This skin also appeared to have a medicinal use: if a reaper got back pains, it was thought that he was being punished by the corn spirit, so he wore the

Thor and his goats.

The goat Heiðrún consumes the foliage of the tree Læraðr, while her udders vent mead into an urn.

goatskin cloak to heal the pain. These practices, although they varied greatly in the precise details, were widespread across the whole of Northern Europe, and in some societies the rituals persisted well into the twentieth century.[24]

Belief in the corn spirit is part of a fertility-cult belief system in which it is important to take proper care of the corn spirit at harvest time in order to ensure that there is a harvest the following year, although the corn spirit itself is not a deity and is not worshipped. There are other fertility cults that include the worship of a goat image. For example, a fertility cult is reported to have existed in the Canary Islands in the fourteenth century. A temple contained sculptures of a woman and a goat engaged in the sexual act, and these were worshipped with milk used as the offering.[25] In the Museum of the Canary Islands there used to be a wooden bowl which, on the upper side, appears as a cross-section through a goat's udder. On the underside, however, the shape suggests a woman's lower body, with a slit cut to suggest the genitals. This seems to further link fertility with goat milk.[26]

In Norse mythology goats have a purely utilitarian role, but with supernatural powers. They feature both as the animals that pull the chariot of the god Thor, and as the animal that provides sustenance for the warriors in Valhalla, the Norse equivalent of heaven. Thor's two goats are called Tanngrisnir (snarler or teeth barer) and Tanngnjóstr (teeth grinder). By day they pull Thor's chariot, but then he cooks them and they feed him and his guests. However, after the meals he gathers the bones of the goats in the goatskins and the next day uses his hammer to bring the goats back to life. There is a story in the Norse legends that one day Thor was staying with a poor farmer and his family, with whom he shared his evening meal. One of the children broke a bone to suck out the bone marrow. From that day onwards one of the goats was lame.[27] In Valhalla there is a female goat, called

Heidrún, who feeds on a tree called Læraör. Instead of producing milk, the teats of Heidrún produce mead that flows into a cauldron, from which the warriors drink their fill.[28]

The use of goats for their milk also has a place in classical Greek mythology, according to which the god Zeus was suckled by Amaltheia, the goat nymph. Zeus was so grateful to Amaltheia that he took her image and set it among the stars as Capricorn. The horn of Amalthea, borrowed by Zeus, is known as the cornucopia, the horn of plenty, which is always filled with whatever its owner desires, symbolizing an abundance of food, drink and wealth. Amaltheia's hide was also used to make a coat for Zeus' son Zagreus.[29]

In marked contrast to the Norse mythology and the legend of Amaltheia, where goats are providers of sustenance, in Judaeo-Christian culture goats are generally used to represent wrongdoing. They are the embodiment of sin and sinners. In traditional Jewish

Robert Lentz, *The Good Shepherd*. This modern Christian icon makes the point that Christ tends to goats as well as sheep (with which he is more usually depicted).

practice, on Yom Kippur, the Day of Atonement, the rabbi takes two goats. The first goat is sacrificed as an offering to God. The second is designated 'for Azazel', which is the origin of the word scapegoat. This goat is given all the sins of the people and sent out into the desert as the scapegoat, taking the sins with it. In Talmudic times the goat was taken to the top of a precipice near Jerusalem and thrown off it. If the tuft of red wool tied to the goat turned white, then God had forgiven the sins of the people.[30] In a similar practice in Tibet, a person was dressed in a goatskin and heard the confessions of the people before being cast out of the community.[31] This powerful concept of the scapegoat was historically very widespread, often involving a human scapegoat who would be sacrificed.[32] The idea persists today, even in a very diluted form, and we still use the word scapegoat to denote somebody who takes the blame for a wrongdoing that was not necessarily theirs.

In a story told by Jesus in the Gospel of Matthew, at the Day of Judgment the people of the world will be divided into the sheep and goats. The sheep are told that they will be handsomely rewarded for having behaved in a charitable manner towards others, providing food and clothing for the hungry and naked, and visiting the sick and those in prison. The goats, on the other hand, will be condemned to eternal damnation for having failed to do these things (25:31–46). The use of sheep and goats in this story probably reflects the herdsman's view of these animals, sheep being rather more compliant and easy to tend than the stubborn goats, which are apt to wander off away from the flock. The formal Christian Latin Mass for the Dead includes the plea 'Among the sheep set me a place and separate me from the goats'.[33] Robert Lentz, the American Franciscan Friar and painter of Christian icons, has painted *The Good Shepherd*, showing Christ tending a billy goat instead of the more traditional lamb, presenting a modern theological image of Christ the saviour.

A legend from Northern Europe supports the Christian idea of goats as wrongdoers. In Finland it is told that a goat was present in the stable at the birth of Jesus. As it was very cold, his mother, Mary, asked the goat for some of its fleece to keep baby Jesus warm, but the goat refused. As a punishment the goat was told by an angel that he had to walk around the world on every Christmas night carrying gifts for little children.[34] This story is one explanation of the origin of the Yule goat. Although the tradition of the Yule goat is observed across Scandinavia, different stories accompany the tradition. It is likely that it started in pre-Christian Europe during the festival of Yule, which involved the slaughter of a goat. Now the goat is usually represented by a straw

Swedish goat as a Christmas decoration.

effigy, which is used to decorate Christmas trees in Sweden and other Scandinavian countries. The use of straw very clearly links the Yule goat back to the tradition of the corn spirit as a goat. In some parts of Sweden a man would be dressed from head to foot in straw, with a pair of horns on his head, and would be led around as the personification of the Yule goat. An extension of this is the Yule play that is enacted in Scandinavia. A man dressed in goatskins and bearing horns is led into a room by two other men, who enact the slaughter of the goat, then sing a song whose verses refer to each colour of the cloths which they lay on him. At the end of the song the Yule goat, having pretended to be dead, suddenly comes back to life, jumping up and dancing around for the entertainment of the audience.[35]

Goats have been used as sacrificial animals throughout the history of mankind. In Judaism goats are acceptable animals for sacrifice, and it was often the case that the flesh of the sacrificial animal was distributed and eaten by the worshippers. In Leviticus a male goat was specified as an appropriate sin offering for a community leader, while a she-goat was acceptable for an ordinary member of the community (4:22–31). Goats are cited throughout the Pentateuch as sacrificial offerings: when Moses set up the altar to the Lord, each of the twelve tribes of Israel brought sacrificial animals, which included a total of 60 goats (Numbers 7).

The ancient region of Kafiristan (now part of Afghanistan) had a law that specified that at the second principal feast of the year, each family must sacrifice a goat.[36] In the Kalasha culture in northern Pakistan goats are sacrificed both on important religious occasions, and particularly at the funerals of both men and women.[37]

From Roman times goats have had strange, almost supernatural characteristics attributed to them. The Nubian ibex, according to reports of Pliny's writing, could run at great speed

John Bauer, *Julboken* or Swedish Yule goat, 1912.

and throw itself from great heights, landing on its horns, which were elastic and functioned as shock absorbers, therefore enabling the ibex to bounce when it landed.[38] What Pliny actually said was:

> There are the capræa, the rupicapra or rock-goat, and the ibex, an animal of wonderful swiftness, although its head is loaded with immense horns, which bear a strong resemblance to the sheath of a sword. By means of these horns the animal balances itself, when it darts along the rocks,

Nicolas Poussin, *The Triumph of Pan*, 1636.

as though it had been hurled from a sling; more especially when it wishes to leap from one eminence to another.[39]

In such a way are mythologies built. Pliny also believed that goats had such excellent night vision that anybody who drank goat's blood or ate goat liver could acquire the ability to see in the dark.[40] Goats could also use their horns for breathing, according to Oppian. It was not clear how the hornless goats described in Roman times might breathe, although Pliny quotes Archelaus as believing that goats breathed through their ears.[41] It was also thought that the teeth and saliva of goats were poisonous to all vegetation: Cosinius argued that a goat could make a vine whither or an olive tree sterile by simply eating a small leaf, while Pliny believed that goats made an olive tree barren by licking it.[42]

Pliny quotes a story, told by Mutianus, of two goats crossing a bridge, which is used to illustrate the exceptional intelligence of the goat:

> Two goats coming from opposite directions met on a very narrow bridge, which would not admit of either of them turning round, and in consequence of its great length they could not safely go backwards, there being no sure footing on account of its narrowness, while at the same time an impetuous torrent was rushing rapidly beneath; accordingly one of the animals lay down flat, and the other walked over it.[43]

In medieval England, as in ancient Rome, goats were seen as creatures with special powers. The bestiaries written in England in the twelfth and thirteenth centuries give an insight into how both real and mythical animals were seen in the medieval world. According to these writings goats lived on mountains or high hilltops, and could tell from far off whether a man was a hunter or a traveller. In this respect they were considered to represent the all-knowing and all-seeing vision of Christ himself. It was also thought that goats could distinguish good plants from harmful ones, and it was reported that a wounded goat would find the herb dittany and be healed simply by touching the plant.[44] The bestiaries also added to the sexual reputation of goats, reporting that male goats were 'stubborn and lascivious animals, always ready to mate', and had such a hot nature that the blood of a goat could dissolve diamonds.[45] In the sixteenth century it was believed that if goats lost their balance when climbing rocks, they would somehow 'thrust out their heads against the rockes and hang by their hornes until they have recovered'.[46]

Goats are often associated in some way with evil or ill-omen. They were traditionally considered to be one of 'the Devil's animals'. There was a superstitious belief in parts of rural England

Andrea Mantegna,
*The Triumph of the
Virtues*, 1502.

Crisme Vauderye,
Witches' Sabbath,
1460.

that 'no he-goat ever remains in sight for twenty-four consecutive hours because once a day it pays a visit to the Devil to have its beard combed.'[47] In the Arcadian fertility cult, which is very similar to the European witch cult, the Devil was represented by the half-goat god Pan.[48] In Christian culture the Devil is strongly associated with the goat. In the Mantegna painting *The Triumph of the Virtues* (also called *Minerva Expelling the Vices from the Garden of Virtue*) from 1502, the Vices are depicted as goat-like men, with horns, hairy legs, hooves and an erect penis. The word Azazel (the scapegoat of Hebrew tradition) has pre-Hebrew origins, associated with the worship of demons who took the form of a goat. Later, Azazel was used as the name of a fallen angel, possibly also referring to Satan.[49]

In Tibetan Buddhist mythology the hero Gesar kills, over a seven-day period, a number of demons and animals of ill-omen that include, on the sixth day, a black she-goat.[50] The god

Baphomet, of Aleister Crowley's paganism, is half-human and half-goat, bearing up to three heads in the form of a goat head with horns. Baphomet is the basis for the modern image of Satan and is invoked in black magic rituals, where the head of a goat is placed on an altar.[51] In the incantation recited by the three witches of *Macbeth*, 'gall of goat' is one of the items of ill-omen listed in the recipe for their evil potion.[52] Witches are often depicted riding on a goat instead of a broomstick. Curiously they appear riding backwards on the goat, facing the tail. In medieval times it was widely believed that goats whispered lewd thoughts into the ears of saints.

There are some beliefs, however, that associate goats with the removal of evil or harm, perhaps as a reflection of the scapegoat tradition. For example, in Scotland goats were thought to frighten off the goblin that frequented hillsides, bringing sickness with him. It was also believed in many areas that goats could keep snakes away. The wild goat name 'markhor' means snake eater, reflecting the belief in Kashmir that these goats

A Mughal jade cup in the form of a goat, made in the 17th century.

consumed snakes. Similar ideas were found in rural England where goats were said to kill adders, the only venomous snakes in England, by scooping them up with their horns, trampling them underfoot and then eating their bodies (except for the head). There is an old English saying to the effect that 'There's no luck where there's no goat'.[53] The non-Muslim Kalasha people of northwest Pakistan regard the ibex as a sacred animal. They believe that the ibex herds belong to the gods who live in the high mountains. Their domestic goats are also highly regarded because they are seen as pure animals that are acceptable for religious sacrifice, so are tended accordingly.[54]

Throughout Afghanistan and northern Pakistan local mosques are decorated with ibex horns and some Muslim groups, particularly the Chitrali Muslims, hold the ibex in especial esteem. The meat is regarded very highly both because of its inherent healthful qualities, as the animals feed only on high-mountain plants, and because it is believed not to shrink when cooked as other meats do; instead it increases in size and so becomes more plentiful. In a variant on the polytheistic views of the Kalasha people, the Chitrali believe that the ibex belong to mountain spirits who protect the goats. There is a story that one man in the region who had killed more than 300 ibex had become bald, which was uncommon in that district. His neighbours said that the spirits had taken his hair to punish him because he had taken their goats.[55]

There are many stories in which animals, frequently including goats, are given human characteristics and used to illustrate either wise or foolish actions. The most famous of these are *Aesop's Fables*. In this collection of morality tales attributed to the Greek slave Aesop, who lived around 620–560 BC, goats have a mixed press. In the story 'The Fox and the Goat', the fox takes advantage of the goat's rather innocent and trusting nature:

A Limoges roundel, *The Temptation or The Fall*. Eve looks over her shoulder at a malevolent-seeming goat in a representation of temptation.

A fox fell into a well and couldn't escape. Soon a goat came along and, seeing the fox in the well, asked if the water was sweet. The fox assured him that the water was marvellous and suggested that he jumped into the well to try it for himself. As soon as the goat was in the well the fox pointed out that they were now both stuck and suggested a way of escape. He proposed that the goat should stand on his hind legs bracing his horns against the wall and allow the fox to run up his back and out of the well. The fox would then help the goat out. The goat readily agreed to this plan and soon the fox was out of the well. As he ran away he called back at the goat, still stuck in the well, 'You foolish old goat. You should have been more careful to make sure that you could get out of the well before jumping in.' The moral of the story is 'Look before you leap.'

In other fables the goat is rather more worldly-wise. In 'The Goat and the Goatherd', the goatherd is trying to bring a stray goat back into the flock. He calls and whistles but the goat pays no attention. In the end he throws a rock at the goat, which breaks its horn. The goatherd begs the goat not to tell anybody what he has done, but the goat replies that even though he will say nothing, his broken horn will speak for itself.

The traditional folktales of Northern Europe portray goats as wise animals that outwit those who wish them harm. The story of the goat and the seven little kids tells of how a mother goat leaves her kids while she goes looking for food. When she returns she finds that they have all been eaten by a wolf, which is now asleep under a tree. The mother goat cuts the wolf open, releasing her kids, which have all been swallowed alive. She replaces

The biggest and fiercest Billy Goat Gruff.

Statue of the Five
Goats, Guangzhou
province, China.

Statue of the Five Goats, Guangzhou province, China.

the kids with big stones that she sews into the wolf's stomach. When the wolf awakes he is thirsty and goes looking for a drink. When he bends over to drink from a well the stones cause him to overbalance and fall in the well, where he drowns.

In other stories goats use guile combined with force. One such story has several forms around Northern Europe, but is best known as 'Three Billy Goats Gruff' in the Grimm Brothers' collection of fairy tales. There are three goats who wish to cross a river to reach the meadows on the other side, but the bridge is the home of a fierce troll. They approach the bridge one at a time. The first and smallest goat tells the troll to wait for his larger, and tastier, brother and is allowed to cross. The second goat makes the same suggestion and is also allowed to cross. The third goat, however, is even larger and fiercer than the troll and charges the troll, knocking him into the river.

A story from Central Africa explains how goats came to have such a strong and unpleasant odour. Goats are particularly numerous in Central Africa, and apparently they had gone to the local goddess and asked if they could share her box of perfumed ointment so that they might smell as fragrant as her. The goddess was outraged at the request and, taking a different box of ointment, with a most unpleasant small, anointed the goats with this noxious ointment instead of the fragrant one. From that day on goats have smelled bad to remind them of their punishment for being impertinent.

Guangzhou, a city in China, has a statue of five goats that mark the legendary founding of the city. According to tradition, five celestial beings came to the city, each riding a five-coloured goat and carrying an ear of corn. They gave the ears of corn to the inhabitants of the city with the promise that there would never be famine in Guangzhou. The city is also known as Wuyang City, meaning five-goat city.[56]

Goats thus have a place in the myths and folktales of peoples across Europe and Asia. In our mythology goats are lustful and promiscuous, yet nurturing; they are both foolish and wise; they represent Satan and evil, yet protect from harm. They are honoured and made scapegoats, worshipped as sacred and sacrificed to gods. They are believed to have supernatural powers for both good and ill, yet they are wholly commonplace.

5 The Medicinal Goat

If the beard were all, the goat might preach.
English proverb

Since the time of ancient Egypt goats and their products have been believed to have various curative properties. While the specific beliefs have altered with time, there remains a strand of thought that holds the goat and its products as having some special medicinal value. At different times and in different cultures, every single part of the goat has been used for its health-giving effects. A wide spectrum of beliefs remains in the world in the twenty-first century.

The earliest reports suggest that goat fat was used in ancient Egypt as a cure for hair loss, but when Aristotle started losing his hair he rubbed goat urine on his head.[1] Pliny records, throughout his *Natural History*, the extraordinarily wide range of medicinal uses that the Romans had for goats. For example, the legs of locusts mixed with suet from male goats was a cure for leprosy,[2] while ox-gall mixed with goat urine was considered an effective treatment both for 'hardness of hearing' and for 'suppuration'.[3] As a 'remedy for pains in the neck' Pliny recommends goat urine again, but this time injected into the ears.[4] For the treatment of snake bites almost any goat product could be used:

> The fumes of the burning horns or hair of a she-goat will repel serpents, they say: the ashes, too, of the horns, used either internally or externally, are thought to be an antidote

to their poison. A similar effect is attributed to goats' milk, taken with Taminian grapes; to the urine of those animals, taken with squill vinegar; to goats' milk cheese, applied with origanum; and to goat suet, used with wax.[5]

Gargoyle
at Avranches
Cathedral, France.

Goatskin and dung were also thought to be effective if applied to the site of a snake bite, with wild goats being more potent than domesticated ones, and males more so than females. Pliny did note that the use of the goat in any medicinal way was a surprise to him, as 'the goat is never entirely free from fever'.[6]

An Anglo-Saxon remedy from *The Lacnunga* 'in the event of blood welling through a person's mouth' was to 'take the weight of three drachmas of betony and three cupfuls of cool goats milk and let him drink it; then he will soon be well'.[7] It is very difficult to imagine any condition causing blood to well through a person's mouth that might conceivably be cured by herbs and milk. For 'tightness of the chest' the cure was to 'boil holly-bark

in goat's milk and eat warm while fasting'. In Iranian folklore
the skin of a blue goat was used to cure blindness.[8] In the early
eighteenth century it was recommended that goat hair or horn
be burned in order to raise somebody out of a deep sleep. Belief
in the curative powers of goats has persisted. As recently as 1945
it was reported that goat fat was used as a salve, rubbed into the
neck and chest, in order to cure the common cold.[9]

Perhaps the strangest and most interesting medicinal association with goats is the bezoar stone of medieval medicine. The
Persian wild goat (*Capra aegagrus*) has the common name bezoar,
which derives from the Persian word *padzahr*. This animal was
the original source of the bezoar stone, or goat stone, which was
believed to have almost magical healing properties. Belief in the
medicinal power of the bezoar originated in medieval Persia and
spread around the world, persisting into the twentieth century.
Indeed, by the seventeenth century *padzahr* had come to mean
antidote. In the USA bezoars are known as madstones.[10]

A bezoar or bezoar stone is a solid concretion found in the
digestive systems of various animals, but most commonly in
those of ruminants such as goats. The stones are formed from
hair and vegetable matter, hardened by digestive juices, and are
often made up of concentric layers like those of an onion. They
are not passed by an animal but remain within the stomach,
often for years, growing in size.

Bezoar stones were once so highly prized that a market in fake
stones was established, with a mixture of wax and herbs being
used to create the fakes. There are passing references to bezoar
stones in tenth- and twelfth-century writings, but the earliest
work devoted to the subject dates from sixteenth-century Persia.[11]
In England the bezoar stone was apparently used to cure King
Edward, although it is not clear which one, and it is also thought
that Elizabeth I wore a bezoar stone set into a ring. The bezoar

Bezoar stones.

certainly featured in seventeenth-century pharmacopoeias, along with hart's horn and other remedies.

It was believed that the bezoar was an antidote to poison of all types. It could be applied topically to a wound, or powdered and given orally. In the sixteenth-century treatise of the Persian doctor Imad-ul-din, it was suggested that the stone was effective against hornet and scorpion stings, and snake bite.[12] It was reported to be effective if ground up and used as a poultice to extract stings, or if taken orally dissolved in wine or milk. The bezoar dose was calculated by weight equivalent to grains of barley: twelve grains to treat bites or stings, one grain to treat weakness of the heart or loss of sexual prowess. It was believed that a dose of a grain a day would preserve the person from all illnesses and poisons. However, it should only be taken for five days in every seven, and when swallowed should not touch the teeth as it would cause them to break. It is easy to see how people in the medieval world, with a view of medicine that held that much illness was caused by poison in some form, whether from stings or infections, would readily believe in this supposed remedy.

It is perhaps surprising that the efficacy of the bezoar was questioned so early, in fact at the same time that Doctor Imad-ul-din was preparing his treatise. Ambroise Paré, the sixteenth-century French surgeon, conducted an experiment to test the bezoar stone as an antidote to poison. Charles ix asked Paré about the effectiveness of the bezoar, and it was agreed that it should be tested on a cook who had been found guilty of stealing his master's silver and had been sentenced to death by hanging. The cook agreed to take part in the experiment on the understanding that if the bezoar proved effective, he would be given his freedom. The cook was duly given 'a certain poison' and a concoction of the bezoar stone was also administered. He died 'miserably' seven hours later, thus proving that the bezoar stone was not a universal antidote to all poisons.[13]

In 1715, around 150 years after Paré, Frederick Slare presented a paper to the Royal Society of London in which he described a number of experiments he had carried out upon his patients, in which bezoar stones had proved to be entirely without use compared with the remedies he usually employed.[14] The experiments

Marginal decoration from *The Romance of Alexander* (1338–44).

were not quite as brutal as the test conducted by Paré in the French court, but were rather more extensive. Slare's conclusions about the value of the bezoar were:

Third-century pavement mosaic of an ibex with a phoenix.

> Nor do I think, that if this dull Idol, which in our Materia Medica has been admir'd and ador'd and to whom such Immense Sums of Money have been ridiculously offer'd for so many Ages, were tumbled down and buried never to rise and more, that Mankind would suffer any Loss or Injury.

It might be thought that such eloquent debunking of the bezoar-stone myth might have put the matter to rest. However, there are reports of bezoar stones (madstones) being used throughout the eighteenth and nineteenth centuries in the USA, and in 1913 a local newspaper in Tennessee offered for sale a madstone with a 'proven record of cures'.[15] Most recently the bezoar stone has reappeared in the Harry Potter books, where it is an essential ingredient in an antidote in potions classes and saves Ron Weasley's life when he has taken poisoned mead.[16]

It is not just the bezoar that was believed to have medicinal uses. In Austria in the seventeenth century, Prince-Archbishop Guidobald Graf von Thun (who ruled Salzberg in 1654–68) established an 'ibex pharmacy' as part of the royal pharmacy in Salzberg, which caused a huge increase in the value of the ibex and consequently a great increase in the numbers of ibex killed.[17] Virtually every part of the ibex's body was thought to have some medicinal value: the blood was used to treat bladder stones, as an antidote to poison and as a treatment for fevers. Ibex dung was a treatment for diseases ranging from tuberculosis to sciatica, gout and arthritis, while the horn was boiled in milk and used for 'female complaints'. It is likely that the only thing that use of this ibex 'medicine' achieved was the complete extermination of the ibex in Switzerland and much of Europe by the start of the eighteenth century.

Sometimes just the presence of the goat was thought to have curative properties. In parts of England it was the custom to bring a goat into the house where somebody was ill in the belief that the illness would leave the person and be taken away from the house by the goat.[18] Goats were also apparently kept in the nursery to keep sickness away from children. These traditions seem to relate to the concept of a goat as a scapegoat. The Kaffirs of southern Africa would take a goat into the house of a sick man, confess sins over the goat and sprinkle the animal with some blood from the sick man. In this way the sickness was transferred to the goat, which was set loose into the bush.[19] Similarly the medicine man of the Baganda people would 'pass some herbs over the sick man's

Charm made of goatskin, found in Nigeria in the early 20th century.

body', then tie the herbs to a goat and drive it out onto wasteland in order to remove an illness from a patient.

There have also been reports of veterinary uses of goats, for maintaining even a small number of goats on a farm has been found to be beneficial to the health of other livestock. In the late nineteenth century it was believed that keeping goats in the stables with horses could prevent the illness known as staggers, a 'nervous disorder' that caused the deaths of many horses. The presence of goats in the same grazing pastures as dairy cattle was also thought to prevent spontaneous abortion in cattle. One school of thought held that the powerful odour of male goats acted to repel the contagion in cattle and to strengthen the constitution of horses. It was, however, also acknowledged that the goat may preferentially consume some plant that caused abortion in the cattle.[20]

Goats still have a significant role in the traditional medicine practised in many parts of the world today. While goats do not have a large role in Chinese medicine, in traditional Tibetan medicine eating goat meat is believed to be beneficial in curing

syphilis, smallpox and burns, as well as reducing fever.[21] Ayurvedic
medicine, a form of Indian traditional medicine that is still practised in many rural areas and is currently gaining popularity in
India, employs rather more unusual goat products, including
urine, faeces and bones. Goat's urine, called *ajamutra*, has been
used to treat a range of disorders, and is either used to form a paste
that is applied to various parts of the body, or even swallowed, to
treat coughs, parasite infections and stomach upsets, while goat
milk and even blood are used to treat other illnesses.[22] In parts of
Kerala province goat urine is drunk to treat tuberculosis, and goat

milk is used for eye disorders. Extracts of goat bones, specifically the thoracic vertebrae, and bone marrow are also used to treat tonsillitis and laryngitis.[23]

Chinese coin.

Although goats are not common in Chhattisgarh, a state in central India, goat faeces are used by traditional healers in the area to treat several ailments.[24] For boils on the head it is stipulated that the faeces must be fresh and applied directly to a boil in order to promote healing. Shoulder stiffness from over-exertion is treated in a similar way, with the fresh faeces being mixed with a number of herbs and formed into a paste with castor oil before being applied to the shoulder. In the same region dried goat faeces are placed in a cloth bag, which is inserted into the vagina as a remedy against excessive menstrual bleeding. For earache the dried faeces are mixed with neem juice (an extract of the Indian lilac tree) and used as ear-drops. At least none of these remedies involves swallowing the faeces: the excrement of some other animals is thought to be most effective if eaten. In Arunchal Pradesh in north-east India, the frontal bone of the goat's skull is burned and the ash taken in pinches mixed with water several times a day. It is thought to both cure fever and relieve any pain in early pregnancy. The gall bladder of the goat is cooked with rice and used to treat stomach ache.[25] In some parts of South America there is also a medicinal use for goat parts; goat horn is used to treat 'ribcage pain', and other goat products are used against 'evil eye, snake bite and muscle strain'.[26] The horn of the Nubian ibex is still sold as a medicine in the markets of Jordan, but not Israel, and its medicinal use is not clear.[27]

In traditional Bedouin society goats are used to protect against the bite of a particular spider, called *ankabŭt*.[28] According to legend this spider acquired the gift of speech and swore that anybody it bit would die. In order to counteract this curse it is necessary for the victim's family to take a neighbour's goat, slaughter it and

Goat-shaped
amphora from
Cyprus, 600 BC.

place its entrails in water. The bite is then washed with this water, but the victim must also sit in a freshly dug grave for a few moments to completely break the spell.

Goat milk has a far lower 'yuk' factor than urine, faeces and horn, and it is commonly used in Ayurvedic medicine. The Hebrew Talmud suggests that milk fresh from a goat will relieve heart pain and, furthermore, that milk from a white goat has special curative properties.[29] The Garasiya people of Rajasthan use goat milk for treating both mouth ulcers and asthma.[30] However, it is in alternative forms of Western medicine where goat milk has been the focus of attention and the subject of fierce debate. The debate was opened by Thomas Bewick at the end of the eighteenth century, who first reported that

> The milk of the goat is sweet, nourishing and medicinal, being found highly beneficial in consumptive cases. It is not so apt to curdle on the stomach as that of a cow . . . It is made into whey for those whose digestion is too weak to bear it in its primitive state.[31]

The claim for its use in the treatment of 'consumptive cases', now known as tuberculosis, has not been repeated, but since Bewick's time various claims have been made for the health benefits of goat milk. Over the past three decades goat milk has been promoted in the developed world as a healthy alternative to cow milk. While the debate as to whether goat milk is inherently better or worse than cow milk will probably never be entirely resolved, some of the pro-goat-milk campaigns have caused real health problems.

The original suggestion that goat milk might be used to feed infants was made in the eighteenth century: goats were cheaper than human wet nurses and easier to obtain, and it was suggested that infants might suckle directly from a goat as the act of sucking was thought to be beneficial.[32] In 1816 a book was published called *The Goat as the Best and Most Agreeable Wet Nurse*, by Conrad Zweirlein, in response to him hearing a group of fashionable young women at a health resort complaining about the difficulty of finding a wet nurse. The use of goat milk became briefly popular in Europe, but was always controversial, partly because it was feared that children would acquire the 'libidinous nature of goats',[33] and it fell out of fashion until the late twentieth century.

Goat monument, Uryupinsk, Russia.

In the 1970s, advice from the British Goat Society, which was widely reproduced elsewhere, suggested that babies and infants should be fed goat milk.[34] Similar advice has been given in some alternative health forums, suggesting that goat milk can be used in place of breast milk or proprietary infant formula.[35] This advice seems to be based on a serious misunderstanding of the role of animal milk in infant-milk formulae. There can be no doubt that unmodified milk, whether from cow or goat, is not a suitable feed for infants. Goat milk has both a high salt concentration and a very low concentration of folate, which can make infants very unwell.[36] The medical literature describes the use of unmodified goat milk for infant feed as a 'dangerous practice'.[37]

Even modified milk can cause problems. An 'alternative' recipe for a homemade infant-feed formula suggests a mixture of whole goat milk, barley water and corn syrup. This recipe was published in 1972 in a book by Adelle Davis called *Let's Have Healthy Children*. As a result of following the advice in this book, a number of children suffered serious health problems, including severe goat-milk anaemia, caused by the low folate levels.[38] The book was withdrawn from sale and the author's estate was successfully sued for the damage her book caused. However, the advice she gave is still repeated on various websites, and secondhand copies of the book are still readily available. Perhaps it should have been renamed 'Let's Have Unhealthy Children'.

There are several proprietary versions of infant formula based on goat milk on the market, and like any other infant-formula food these have been properly supplemented and adjusted to form a safe and properly balanced substitute for breast milk. Whether a goat-milk formula is inherently any better than a formula based on cow milk is a matter of personal opinion. There is certainly no scientific evidence for it. What is certain, however, is that a homemade formula of goat milk, whatever it is modified with, is not an appropriate food for a baby.

Goat milk is popularly believed to be an effective treatment for infant eczema, and to be more digestible than cow milk. There is little scientific evidence to support either of these claims. However, the effect of the alternative-health movement on the commercial production of goat milk is impressive. Goat milk is now very widely available in supermarkets, not just in health-food shops, and this interest has surely contributed to the increased interest in goat-keeping across the developed world.

In the developing world there is a very strong case for the health benefits of goat milk. In areas of the world where food is in short supply, the presence of a family goat and consumption of its milk can have a hugely beneficial impact on the family. This impact is particularly seen in relation to the health and development of children who may otherwise be seriously under-nourished. In these circumstances goat milk is beyond doubt of outstanding health value, and can make the difference between life and death for a child.

6 The Twenty-first-century Goat

If the goat had a longer tail he could wipe the stars clean.
Czech proverb

There is still the belief in some communities that goats have supernatural powers. In 2009 there was a news report of a goat being arrested on suspicion of burglary in Nigeria. The goat was accused of car theft and of using witchcraft to shape shift.[1] In 2010 world news agencies ran stories about the president of Pakistan, Alif Al Zardari, using the almost daily ritual slaughter of goats to ward off evil eyes and fend off black magic. A spokesman for President Al Zadari denied that this was the primary purpose of the sacrifice: 'The main belief is that this practice invokes the pleasure of God. The corollary is that bad things will not happen but that's a matter of interpretation.'[2]

Even in the developed world there are some odd beliefs about goats. In 2006 a goat was demoted in rank in the British Army for 'inappropriate behaviour' during the Queen's birthday celebrations. William Windsor, or Billy the Goat, is the name given to the regimental goat of the 1st Battalion of the Royal Welsh Regiment, in the British Army.[3] The goat is not considered to be a mascot, but a serving member of the battalion, usually admitted as a fusilier, but sometimes reaching the rank of lance corporal. The regimental goat is a tradition that dates back to the Crimean War, but the first 'official' goat was presented to the Regiment by Queen Victoria in 1862. Since then goats have seen frontline action with the Welsh Regiment, most notably during

The goat presented by the king to the 7th Royal Welch Fusiliers, 1914.

the First World War, when a goat called Taffy IV was awarded a number of battle honours.[4] The goat wears a silver headplate that identifies it as a gift from the monarch. The regimental goat is a cashmere goat, chosen from either the herd at Whipsnade Zoo or the herd on the Great Orme in Llandudno, North Wales.[5]

The 'Curse of the Billy Goat' is blamed for the lack of sporting success of the Chicago Cubs baseball team.[6] The Cubs have failed to achieve major honours since 1945. According to legend the owner of the Billy Goat Tavern was asked to leave the home baseball game as he smelled strongly of goats, and as he left he placed a curse on the ground. Various attempts have been made to break the curse, usually involving fans bringing goats along to a home game, but the ground has also been blessed by priests and exorcised. While the Cubs have won their division, they have still never made it to the World Series.

Goats are occasionally used rather more positively, as mascots for sporting teams and other organizations. The image of a goat called Fitzbilly is the mascot of Fitzwilliam College, Cambridge University. The Royal Air Force Aircraft Apprentices used to have goats as their mascots, and now E-Goat is the name of the 'unofficial RAF rumour network'.[7] Cologne Football Club, which plays in the German Bundesliga, has a goat mascot called Hennes. The original Hennes was given to the club by a circus owner in 1960, and today Hennes VII appears at all the club's home games. His fans can keep an eye on him through his very own webcam.[8]

Bill the Goat is the official mascot of the United States Naval Academy.[9] Goats were used on board naval ships to provide food for the sailors and were often regarded as pets. U.S. naval legend tells that when one much-loved goat died on board a naval ship, its skin was saved to be preserved and mounted back in port. However, on arrival in port the sailors went to watch an American football game in which the Naval Academy was participating, and at half time one of them dressed in the goatskin in order to

U.S. Naval Academy mascot, Bill the Goat.

This goat was ejected from Wrigley Field, Chicago, on 12 October 1945 during the World Series, despite having his own paid seat. The angry owner, William Sianis, cursed the Cubs, saying they would never again win a World Series. They never have.

provide entertainment. The naval victory that day was attributed to the goatskin, and since 1893 a live goat has been adopted as the mascot of the u.s. Naval Academy football team. The goat is particularly significant in the context of the intense rivalry between the Army and Navy. This has resulted in regular kid-napping attempts of the Naval Academy goats by Army cadets, despite high levels of prohibition of such activities. While these goats have had a variety of names in the past, including the name

Buzkashi, Afghanistan's national sport. The name means 'goat-grabbing'!

Satan given to a particularly bad-tempered individual, the great majority have been called Bill, with the present holder of the post being Bill xxxiv.

Goats play a rather different part in some sporting traditions. Buzkashi is Afghanistan's national sport, played on festival days such as Eid and at local carnivals. The name goat-grabbing is said to derive from the times when champion hunters went up into the mountains on horseback to kill wild goats.[10] It is still played on horseback, but now uses the body of either a goat or a calf, usually beheaded. The object of the game is to place the animal's body across a goal line. It is a rowdy, noisy and dangerous sport.

Annual festivals are held to celebrate goats in some parts of the world. In Marshall County, Tennessee, there is a goat festival, started in 2002 and held in October each year, which celebrates

136

all goats, but especially the famous 'fainting' or myotonic goat, which is a breed found only in the USA. This variety of goat was first seen in 1880 when a traveller came to Marshall County with four goats. These animals had a very odd behavioural trait: when startled they had a seizure in which their muscles went into spasm, causing them to fall over. Since then these unusual animals have achieved a sort of cult status and are widely celebrated at goat festivals throughout Tennessee. The Marshall County festival includes a classic goat show for the myotonic breed, as well as music and entertainment.[11] The town of Spindale, North Carolina, started an annual dairy goat festival in 2010. This festival features a goat parade, including goats, carts and their owners dressed in various costumes.[12] An annual goat-racing festival is held in Falmouth, Pennsylvania, each year, which features both goat racing and a goat rodeo.[13] Similar festivals are held in other parts of the USA.

The goat carnival, or Aprokreis, held every spring on the Greek island of Skyros, has a far older and rather more obscure origin than the festivals held in the USA. The men of the island dress in goatskins and perform a goat dance.[14] The carnival procession is accompanied by much eating and drinking. This pagan festival is probably a relic of Dionysian cults from pre-Christian times, and has simply been assimilated into the Christian calendar.

Throughout Scandinavia the Yule goat, or *julebukk*, is as much a part of Christmas as a decorated tree. There was an ancient pagan tradition in Norway of the 'goat', who symbolized the ghosts of winter, going round houses and receiving gifts and bringing gifts to children at Christmas. By welcoming the goat the local people could ward off harm for the coming year. In the twenty-first century it is Santa Claus who brings the gifts, and the children who dress in costume with face masks, and go round the houses singing, and giving and receiving gifts. The goat is now a Christmas tree decoration, made of straw with a piece of red cloth tied to it. A popular custom is to take a *julebukk* into a neighbour's house and conceal it somewhere. When it is found the neighbour repeats the process with another family. In this way the *julebukk* travels from house to house throughout the neighbourhood, bringing the community together in the middle of a dark winter, as it did in pagan times.[15]

In the Swedish city of Gävle there is a Christmas goat tradition that started as recently as 1966.[16] Local businessmen decided to build a giant straw goat in the centre of the city. The first goat was 11 m (36 ft) high, 7 m (23 ft) long and weighed 3 tonnes. Perhaps not surprisingly, it was burned down. Since then a straw goat has been built in Gävle every Christmas, and the local organizers and fire brigade have been trying every means to prevent it from being destroyed. There has been something of a cat-and-mouse game between the organizers and the arsonists, with the organizers managing to prevent more than half the goats from being destroyed. They have used net protection and a fire-retardant spray, and in 1996 introduced a webcam. This has had only limited success, making the arsonists more resourceful. In 2009 a coordinated attack by hackers, who put the webcam out of action, and arsonists, resulted in the burning of the goat. Since 1986 there have been two goats built every Christmas: fans eagerly watch to see whether they survive until New Year!

A tradition of goat-racing has recently become established in several countries. The Caribbean island of Tobago, for example, holds a Buccoo Goat Race Festival every Easter. The races held at this festival originated in 1925 as a cheaper alternative to horse races for the working classes. The 'jockeys' do not ride the goats, but run the 90-m (100-yd) course, dressed in uniform, with their goat on a halter.[17] In Australia goat-races employ feral goats,

The Christmas goat in Gävle, Sweden.

which pull carts around a racing circuit. In London, on the same day as the famous University Boat Race, contested by Oxford and Cambridge University rowing crews on the River Thames, a less famous goat race is held in Spitalfields' City Farm.[18] Two of the farm's goats are selected and each is equipped with a tiny rag-doll 'jockey' made in either the light blue of Cambridge or the dark blue of Oxford. The two goats run a short track through the farm, for the entertainment of the visitors. In the Royal Ascot Goat Races, which have been held in Kampala, Uganda, since 1993, the goats do not have jockeys of either the human or cloth variety, but run the course on their own, cheered by large numbers of spectators, in a charity fundraising event.[19] The tradition of charity goat racing has now spread out from Uganda, through East Africa.

In the early twenty-first century goats became the gift of choice for the fashionably anti-consumerist movement as a wedding or Christmas gift. They were not actual goats, as would

Goat-racing in Australia using feral goats.

be usual in the Hindu Kush, but charity goats. The trend was started by a London-based charity called FARM-Africa, which had been working to promote sustainable agriculture in the developing world by interventions such as providing a goat to a family to enable it to establish some degree of economic independence.[20] A goat provided by the charity featured in a documentary, and on the back of this FARM-Africa launched its 2005 fundraising campaign using the Gruff Family of goats. The campaign was a huge success and by 2006 the idea had been adopted by several major charities.

The idea is that a donor buys a goat for a family in the developing world. The goat provides milk, wool and, of course, more goats, which can all be sold to provide income for the family.

FARM-Africa encourages donors to purchase a goat by making a donation of around £30.[21] The donor receives a gift pack that includes a small plastic goat. Of course, not all of the 'goats' purchased end up as actual goats with a family, as this could result in vast excesses of the animals where they are not really needed. FARM-Africa makes it clear that it uses the donations to support its work in the developing world, which includes community training and education programmes, as well as giving actual chickens or goats to families.

Charity goat.

The charity goat has revolutionized fundraising for work in the developing world. Until 2004 most fundraising depended on images of deprived children. Suddenly there was a positive, cuddly image of the developing world: cute goats. A campaign based on 'adopt a child' for Christmas could never have worked,

Goat-racing in London.

but the charity goat chimed perfectly with the zeitgeist. It was an ideal ethical gift, perfect for last-minute shopping, as highlighted by the high-profile website lastminute.com in its 2005 Christmas gift list. Since 2004 the charity goat has raised something in the region of £10 million for overseas development work.

In North Africa, in particular Morocco, goats have recently become a tourist attraction. Not far from Marrakesh are argan orchards, which are cultivated for the oil obtained from the argan nut. The oil is used in the cosmetics industry and is extremely valuable, although extracting the oil-bearing kernels from the argan fruit is difficult work. It is much easier to do if the nut has first passed through the digestive system of a goat. Goats are allowed into the orchards at harvest time. They climb the trees to reach the fruits, and tourists flock to the orchards to see the mammals in the trees. The phenomenon is so popular that a calendar featuring pictures of goats in trees is published every year.

Goats have also appeared prominently in books and films in the early twenty-first century. Jon Ronson's book *The Men Who Stare at Goats* (2004) tells of the American military's attempt to harness supernatural forces, such as the ability to kill goats simply by staring at them. The beautiful Italian film *Le Quattro Volte* (2010) tells of a goatherd's life in the impoverished area of Calabria. Edward Albee's play *The Goat* (2002) explores questions of sexual behaviour and social constraints, while the film *Goats* (2012), based on Mark Poirier's novel of the same name, is about a teenage boy's relationship with Goat Man, and his coming of age. Suddenly goats have become very mainstream indeed.

Goats thus have an interesting position in the world today. They provide meat, milk and medicine, and are important to the welfare of subsistence farmers in many parts of the world. They are also invasive and are subject to control and eradication, yet

Goats climbing an argan tree in Morocco.

143

Ar lenuue du dyable
la mort print entree
ou monde . Et ce te
enſuiuent ceulx qui tiennent ſō

their wild counterparts are rare and highly protected. We race goats, hold festivals in their honour and parade them as military and sporting mascots, yet we also endow them with supernatural properties and hold them as symbols of both good and evil.

Waldensian Kissing the Hind End of a Goat', a manuscript image ridiculing followers of the radical Waldensian movement, in Johannes Tinctoris' *Sermo contra sectam valdensium*, c. 1450–75.

Timeline of the Goat

8.2 MYA	1.77 MYA	30,000 BC	7000 BC	5000–3000 BC
Ancestral goat, *Pachytragus*, living in Greece and Iran	First true goat living in Georgia	Ibex paintings in Chauvet, France	Domesticated goats in Jericho	Goat images on pottery in China

1200S	1500	1620S	1660S
First association of goats with the Devil	Goats introduced to island of St Helena	Goats introduced to America	Establishment of ibex pharmacy in Salzberg

1893	1904	1945	1970
Goat mascot adopted by U.S. Naval Academy football team	Dairy goats first introduced into USA	Last time the Chicago Cubs win major honours – Curse of the Billy Goat	First goats on the northern part of the Galápagos island of Isabela

| c. 4000 BC | 2500 BC | 2400 BC | AD 1086 |

Goats introduced to
Northern Europe

Goat images on
artefacts in Egypt,
Mesopotamia and
Sumeria

Angora breed
developed in Turkey

Domesday Book
suggests that there
are 80,000 goats
in England

| 1731 | 1780s | 1862 | 1879 |

First attempt to
eradicate goats from
St Helena

Goats introduced to
Australia

First official use of a
regimental goat by
armed forces

British Goat Society
formed

| 1997 | 2002 | 2006 | 2012 |

Goat population in
northern Isabela
estimated at
100,000–
150,000

Feral goat declared
a pest species in
Australia, where it
is raced

Northern Isabela
declared goat free
following eradication
programme

Goats eradicated
on St Helena

References

INTRODUCTION

1 Mollie Panter-Downes, *One Fine Day* (London, 1947), p. 2.
2 W. Maggi, 'Kalasha', in *South Asian Folklore: An Encyclopedia*,
   ed. M. A. Mills, P. J. Claus and S. Diamond (New York, 2003),
   pp. 317–19.
3 Pliny the Elder, *Natural History*, vol. XII, ch. 37.
4 G.F.W. Haenlein, 'Past, Present and Future Perspectives of Small
   Ruminant Dairy Research', *Journal of Dairy Science*, LXXXIV
   (2001), pp. 2097–115.
5 The UN Food and Agricultural Organization's statistical service,
   http://faostat.fao.org.
6 'goat, n.', *OED Online* (accessed June 2011).
7 'nanny-goat, n.' and 'billy-goat, n.', *OED Online* (accessed June 2011).
8 'kid, n.', *OED Online* (accessed June 2011).
9 William Shakespeare, *King Lear*, I. ii. 125.
10 R. W. Holder, *Faber Dictionary of Euphemisms* (London, 1989).
11 Christopher Dyer, 'Alternative Agriculture: Goats in Medieval
   England', in *People, Landscape and Alternative Agriculture. Essays
   for Joan Thirsk*, ed. R. Hoyle (Agricultural History Review
   Supplement Series III, 2004), pp. 20–38.

1 A NATURAL HISTORY OF THE GOAT

1 Manuel Hernández Fernández and Elisabeth S. Vrba, 'A Complete
   Estimate of the Phylogenetic Relationships in Ruminantia:

A Dated Species-level Supertree of the Extant Ruminants',
Biological Reviews, LXXX (2005), pp. 269–302.

2 Alan W. Gentry, 'Caprinae and Hippotragae (Bovidae, Mammalia) in the Upper Miocene', in *Antelopes, Deer and Relatives*, ed. E. S. Vrba and G. B. Schaller (New Haven, CT, 2000), pp. 65–83.

3 Peter Grubb, 'Order Artiodactyla', in *Mammal Species of the World*, 3rd edn, ed. D. E. Wilson and D. M. Reeder (Baltimore, MA, 1993), pp. 700–703.

4 B. Nievergelt, 'Sheep and Goats: Ibexes and Wild Goats', in *Grzimek's Encyclopedia of Mammals*, vol. V, ed. S. P. Parker (New York, 1990), pp. 510–15.

5 Heinrich Mendelssohn, 'Nubian Ibex (*Capra ibex nubiana*)', in *Grzimek's Encyclopedia of Mammals*, vol. V, ed. S. P. Parker (New York, 1990), pp. 525–7.

6 J. L. Long, *Introduced Mammals of the World* (Oxford, 2003), p. 500; L. Baskin and K. Danell, *Ecology of Ungulates: A Handbook of Species in Eastern Europe and Northern and Central Asia* (Heidelberg, 2003), p. 241.

7 D. M. Shackleton and C. C. Shank, 'A Review of the Social Behaviour of Feral and Wild Sheep and Goats', *Journal of Animal Science*, LVII (1984), pp. 500–509.

8 Sue Weaver, *Goats: Small-scale Farming for Profit and Pleasure* (Irvine, CA, 2009).

9 Shackleton and Shank, 'A Review of the Social Behaviour of Feral and Wild Sheep and Goats', pp. 500–509.

10 J. Greyling, 'Applied Reproductive Physiology', in *Goat Science and Production*, ed. Sandra G. Solaiman (Ames, IA, 2010), pp. 139–55.

11 B. Nievergelt, 'Sheep and Goats', pp. 510–15.

12 J. A. Delgadillo, et al., 'Importance of the Signals Provided by the Buck for the Success of the Male Effect in Goats', *Reproduction Nutrition Development*, XLVI (2006), pp. 391–400.

13 P. Weinberg, *Capra caucasica*, in IUCN (2010). IUCN Red List of Threatened Species, version 2010.4, www.iucnredlist.org, accessed 26 April 2011.

14 P. Weinberg, *Capra cylindricornis*, in IUCN (2010). IUCN Red List of Threatened Species, version 2010.4, www.iucnredlist.org, accessed 26 April 2011.

15 R. Valdez, *Capra falconeri*, in IUCN (2010). IUCN Red List of Threatened Species, version 2010.4, www.iucnredlist.org, accessed 25 April 2011.

16 P. Weinberg, et al., *Capra aegagrus*, in IUCN (2010). IUCN Red List of Threatened Species, version 2010.4, www.iucnredlist.org, accessed 26 April 2011.

17 P. U. Alkon, et al., *Capra nubiana*, in IUCN (2010). IUCN Red List of Threatened Species, version 2010.4, www.iucnredlist.org, accessed 23 April 2011.

18 B. Geberemedhin and P. Grubb, *Capra walie*, in IUCN (2010). IUCN Red List of Threatened Species, version 2010.4, www.iucnredlist.org, accessed 25 April 2011.

19 S. Aulagnier, et al., *Capra ibex*, in IUCN (2011). IUCN Red List of Threatened Species, version 2011.1, www.iucnredlist.org, accessed 28 June 2011.

20 Nievergelt, 'Sheep and Goats', pp. 510–15.

21 The UN Food and Agricultural Organization's statistical service, http://faostat.fao.org.

22 Valerie Porter, *Goats of the World* (Ipswich, 1996).

23 Henry Stephen Holmes-Pegler, *The Book of the Goat; Containing Full Particulars of the Various Breeds of Goats and their Profitable Management* (London, 1848).

24 Barbara Vincent, *Farming Meat Goats: Breeding, Production and Marketing* (Collingwood, Victoria, 2005).

25 Frederick E. Zeuner, *A History of Domesticated Animals* (London, 1963), pp. 129–52.

26 Pliny the Elder, *Natural History*, vol. XIII, ch. 76.

27 Porter, *Goats of the World*, p. 5.

28 David Mackenzie, *Goat Husbandry*, 4th edn (London, 1980), pp. 206–9.

29 R. V. Short, J. L. Hamerton, S. A. Grieves and C. E. Pollard, 'An Intersex Goat with a Bilaterally Asymmetrical Reproductive

Tract', *Journal of Reproduction and Fertility*, XVI (1968),
pp. 283–91.

30 M. Soller and H. Angel, 'Polledness and Abnormal Sex Ratios
in Saanen Goats', *Journal of Heredity*, LV (1964), pp. 139–42;
E. Pailhoux, et al., 'Contribution of Domestic Animals to the
Identification of New Genes Involved in Sex Determination',
Journal of Experimental Zoology, CCXC (2001), pp. 700–708.

31 Mackenzie, *Goat Husbandry*, pp. 206–9.

32 Marco Festa-Bianchet and Steeve D. Côté, *Mountain Goats:
Ecology, Behavior and Conservation of an Alpine Ungulate*
(Washington, DC, 2008).

2 THE DOMESTICATED GOAT

1 Barbara Vincent, *Farming Meat Goats: Breeding, Production and
Marketing* (Collingwood, Victoria, 2005), p. 7.

2 Frederick E. Zeuner, *A History of Domesticated Animals* (London,
1963), pp. 129–52.

3 Ibid.

4 H. Fernandez, et al., 'Divergent mtDNA Lineages of Goats in an
Early Neolithic Site, Far From the Initial Domestication Areas',
Proceedings of the National Academy of Sciences, CIII (2006),
pp. 15375–9.

5 Gordon Luikart, et al., 'Multiple Maternal Origins and
Weak Phylogeographic Structure in Domestic Goats',
Proceedings of the National Academy of Science, XCVIII (2001),
pp. 5927–32.

6 D. N. Freedman, ed., *Eerdman's Dictionary of the Bible* (Grand
Rapids, MI, 2000).

7 'The Feral Goat (*Capra hircus*)', Australian government leaflet
(Canberra, 2004).

8 Cecil Jane, trans., *The Voyages of Christopher Columbus* (London,
1930).

9 Sue Weaver, *Goats: Small-scale Farming for Profit and Pleasure*
(Irvine, CA, 2009), pp. 12–14.

10 Joyce E. Salisbury, *The Beast Within: Animals in the Middle Ages* (New York, 1994), p. 15.

11 Christopher Dyer, 'Alternative Agriculture: Goats in Medieval England', in *People, Landscape and Alternative Agriculture: Essays for Joan Thirsk*, ed. R. Hoyle (Agricultural History Review Supplement Series, III, 2004), pp. 20–38.

12 Ibid.

13 G. Kenneth Whitehead, *The Wild Goats of Great Britain and Ireland* (Newton Abbot, 1972).

14 Dyer, 'Alternative Agriculture: Goats in Medieval England', pp. 20–38.

15 Ibid.

16 Whitehead, *The Wild Goats of Great Britain and Ireland*.

17 Stephen J. G. Hall and Juliet Clutton-Brock, *Two Hundred Years British Farm Livestock* (London, 1989), pp. 197–8.

18 David Low, *On the Domesticated Animals of the British Isles* (London, 1845), p. 7.

19 Henry Stephen Holmes-Pegler, *The Book of the Goat; Containing Full Particulars of the Various Breeds of Goats and their Profitable Management*, 4th edn (London, 1909).

20 Whitehead, *The Wild Goats of Great Britain and Ireland*.

21 Richard Bradley, *A Survey of the Ancient Husbandry and Gardening* (1725), p. 371.

22 Thomas Bewick, *A General History of Quadrupeds* (London, 1792), p. 67–71.

23 Henry Stephen Holmes-Pegler, *The Book of the Goat; Containing Full Particulars of the Various Breeds of Goats and their Profitable Management* (London, 1848).

24 Flora Thompson, *Lark Rise to Candleford* (London, 1939).

25 Holmes-Pegler, *The Book of the Goat*, 4th edn.

26 Whitehead, *The Wild Goats of Great Britain and Ireland*, pp. 99–105.

27 Ivor Wynne Jones, *Llandudno: Queen of Welsh Resorts* (Ashbourne, 2002), p. 74.

28 Eve Parry, 'The Great Orme Kashmiri Goats', reproduced from 'Aliens on the Great Orme', www.llandudno.com, accessed 29 September 2013.

29 Hall and Clutton-Brock, *Two Hundred Years of British Farm Livestock*, p. 200.

30 Whitehead, *The Wild Goats of Great Britain and Ireland*, pp. 99–105.

31 Rare Breeds Survival Trust website, 'Bagot' (2014), www.rbst.org.uk.

32 Bagot Goat Society, www.bagotgoats.co.uk.

33 Valerie Porter, *Goats of the World* (Ipswich, 1996), p. 22.

34 Whitehead, *The Wild Goats of Great Britain and Ireland*, pp. 99–105.

35 Porter, *Goats of the World*, p. 22.

36 Personal communication with Rare Breeds Survival Trust.

37 Rare Breeds Survival Trust website, 'Bagot'.

38 David Mackenzie, *Goat Husbandry*, 4th edn (London, 1980), pp. 50–52.

39 W. Ellison, *Marginal Land in Britain* (London, 1953).

40 Mackenzie, *Goat Husbandry*, p. 193.

41 Jill Salmon, *The Goatkeeper's Guide* (Newton Abbot, 1976).

42 Data provided by DEFRA.

43 St Helen's Farm, York. www.sthelensfarm.co.uk.

44 Rosemary Hartley, *Food in England* (London, 1954), pp. 163–6.

45 Dyer, 'Alternative Agriculture: Goats in Medieval England', pp. 20–38.

46 Hartley, *Food in England*, pp. 163–6.

47 *Larousse Gastronomique*, English edn (London, 2001).

48 Ibid.; *The Silver Spoon* (London, 2005), pp. 753–5; Simone Ortega and Ines Ortega, *1080 Recipes* (London, 2007).

49 Hugh Fearnley-Whittingstall, *The River Cottage Meat Book* (London, 2004).

50 *The Silver Spoon*, pp. 753–5.

51 Vincent, *Farming Meat Goats*.

52 Hartley, *Food in England*, pp. 163–6.

53 Vincent, *Farming Meat Goats*.

54 Sandra G. Solaiman, ed., *Goat Science and Production* (Ames, IA, 2010), pp. 139–55.

55 The UN Food and Agricultural Organization's statistical service, http://faostat.fao.org.

56 Ibid.

57 G. F. Haenlein, 'Status and Prospects of the Dairy Goat Industry in the United States', *Journal of Animal Science*, LXXIV (1996), pp. 1173–81.

58 G.F.W. Haenlein, 'Past, Present and Future Perspectives of Small Ruminant Dairy Research', *Journal of Dairy Science*, LXXXIV (2001), pp. 2097–115.

59 St Helen's Farm, York, www.sthelensfarm.co.uk.

60 Dyer, 'Alternative Agriculture: Goats in Medieval England', pp. 20–38.

61 Weaver, *Goats: Small-scale Farming*.

62 Hartley, *Food in England*, pp. 163–6.

63 Bewick, *A General History of Quadrupeds*, pp. 67–71.

64 Salmon, *The Goatkeeper's Guide*.

65 Porter, *Goats of the World*.

66 Ibid.

67 Weaver, *Goats: Small-scale Farming*.

3 THE FERAL GOAT

1 J. L. Long, *Introduced Mammals of the World* (Oxford, 2003), pp. 501–15.

2 'Goat', www.feral.org.au, accessed 12 July 2010.

3 Long, *Introduced Mammals of the World*, p. 504.

4 Q.C.B. Cronk, 'The Decline of the St Helena Ebony *Trochetiopsis melanoxylon*', *Biological Conservation*, 35 (1986), pp. 159–72.

5 Long, *Introduced Mammals of the World*, p. 504.

6 G. K. Bar-Gal et al., 'Genetic Evidence for the Origin of the Agrimi Goat (*Capra aegagrus cretica*)', *Journal of Zoology*, 56 (2002), pp. 369–77.

7 UNESCO–MAB Biosphere Reserves Directory, at www.unesco.org.

8 D. M. Shackleton, *Wild Sheep and Goats and Their Relatives: Status Survey and Conservation Action Plan for Caprinae* (IUCN, 1997).

9 'The feral goat (*Capra hircus*)', Australian government leaflet (Canberra, 2004).

10 The Australian Government Co-Operative Research Centre for Sheep Industry, www.sheepcrc.org.au.

11 The Feral Goat Factsheet (2011), www.environment.gov.au, Canberra.

12 Background Document for the Threat Abatement Plan for Competition and Land Degradation by Unmanaged Goats (2008), www.environment.gov.au.

13 Queensland State Feral Goats Factsheet, www.dpi.qld.gov.au, accessed 12 July 2010.

14 'What to Hunt: Feral Goats', New Zealand Department of Conservation www.doc.govt.nz.

15 'Goat', www.feral.org.au, accessed 12 July 2010.

16 'Project Isabela' and 'Project Pinta', www.galapagos.org.

17 'Judas goat', in 'Control and Eradication of Goats (*Capra hircus*)', www.galapagospark.org, 29 June 2009.

4 THE GOAT OF MYTH AND FOLKLORE

1 H. Hackin, *Asiatic Mythology* (London, 1932).

2 Robert Graves, *The Greek Myths*, vol. I, ch. 75 (Harmondsworth, 1960), pp. 253–4.

3 John H. Rogers, 'Origins of the Ancient Constellations: I The Mesopotamian Traditions', *Journal of the British Astronomical Association*, CVII (1998), pp. 9–28.

4 This is mentioned in Cooper but does not appear in Graves's *Greek Myths*. J. C. Cooper, *Symbolic and Mythological Animals* (London, 1992).

5 Students for the Exploration and Development of Space: an excellent online astronomy resource, http://seds.org/Maps/Stars_en/Fig/capricorn.html.

6 'goat, n.', *OED Online*, June 2011.

7 C. S. Lewis, *The Lion, the Witch and the Wardrobe* (Harmondsworth, 1950), p. 14.

8 Graves, *The Greek Myths*, vol. i, ch. 26, pp. 101–3.

9 H. H. Scullard, *Festivals and Ceremonies of the Roman Republic* (London, 1981).

10 Pliny the Elder, *Natural History*, vol. xiii, ch. 76.

11 Sue Weaver, *Goats: Small-scale Farming for Profit and Pleasure* (Irvine, ca, 2009).

12 Heinrich Mendelssohn, 'Nubian Ibex (*Capra ibex nubiana*)', in *Grzimek's Encyclopedia of Mammals*, vol. v, ed. S. P. Parker (New York, 1990), pp. 525–7.

13 William Blake, *The Marriage of Heaven and Hell* [1790] (1908).

14 'goat, n.', *oed Online*, June 2011.

15 Marina Warner, *From the Beast to the Blonde: On Fairy Tales and their Tellers* (New York, 1994).

16 'goat, n.', *oed Online*, June 2011.

17 Aphrodisiac properties of horny goat weed, www.hornygoatweed.us, accessed 28 September 2011.

18 John Masters, *Bugles and a Tiger: A Personal Adventure* (London, 1956).

19 Graves, *The Greek Myths*, vol. i, ch. 27, pp. 103–11.

20 James George Fraser, *The Golden Bough*, abridged edn (London, 1922), p. 514.

21 Graves, *The Greek Myths*, vol. i, ch. 27, pp. 103–11.

22 Fraser, *The Golden Bough*, p. 610.

23 Ibid.

24 Ibid., pp. 596–600.

25 This is reported by Frederick E. Zeuner, *A History of Domesticated Animals* (London, 1963), p. 152, who quotes 'Bernaldez' as the source. However, no further detail is given and no mention of this observation is made in the major work by Bernaldez, describing the voyages of Christopher Columbus. It has not been possible to verify this report.

26 Zeuner, *A History of Domesticated Animals*, p. 152.

27 Heilan Yvette Grimes, *The Norse Myths* (Boston, ma, 2010), p. 85.

28 Ibid., p. 22.

29 Graves, *The Greek Myths*, vol. i, ch. 7, p. 39.

30 R. J. Zwi Werblowsky and G. Wigoder, *The Oxford Dictionary of the Jewish Religion* (Oxford, 1997).

31 F. Hamel, *Human Animals* (London, 1915).

32 James George Fraser, *The Golden Bough, A New Abridgement* (Oxford, 1994), pp. 591–606.

33 Angela Partington, *Oxford Dictionary of Quotations* (Oxford, 1992), p. 477:14.

34 G. Kenneth Whitehead, *The Wild Goats of Great Britain and Ireland* (Newton Abbot, 1972), p. 59.

35 Fraser, *The Golden Bough*, pp. 552–3.

36 J. Hackin, 'The Mythology of the Kafirs', in *Asiatic Mythology*, ed. H. Hackin (London, 1963).

37 W. Maggi, 'Kalasha', in *South Asian Folklore: An Encyclopedia*, ed. M. A. Mills, P J. Claus and S. Diamond (New York, 2003), pp. 317–19.

38 J. C. Cooper, *Symbolic and Mythological Animals* (London, 1992), p. 137.

39 Pliny the Elder, *Natural History*, vol. XIII, ch. 79.

40 Ibid., ch. 76.

41 Ibid.

42 Quoted in Valerie Porter, *Goats of the World* (Ipswich, 1996), p. 166; Pliny the Elder, *Natural History*, vol. XIII, ch. 76.

43 Ibid.

44 Richard Barber, *Bestiary, MS Bodley 764* (Woodbridge, 1992), pp. 54–7.

45 Ibid.; J. Simpson and S. Roud, *A Dictionary of English Folklore* (Oxford, 2000).

46 Tubervile, quoted by G. Kenneth Whitehead, *The Wild Goats of Great Britain and Ireland* (Newton Abbot, 1972), p. 71.

47 Whitehead, *The Wild Goats of Great Britain and Ireland*, p. 59.

48 Graves, *The Greek Myths* (Harmondsworth, 1960), vol. I, ch. 26, pp. 101–3.

49 Zwi Werblowsky and Wigoder, *The Oxford Dictionary of the Jewish Religion*.

50 H. Hackin, 'The Mythology of Lamism', in *Asiatic Mythology*, ed. H. Hackin (London, 1932), pp. 182–4.

51 J. Ludd, *Secret Art and Magical Practice* (Baphomet Lodge: no city or year given), p. 20.

52 William Shakespeare, *Macbeth*, IV.1.

53 Whitehead, *The Wild Goats of Great Britain and Ireland*, pp. 58–73.

54 *South Asian Folklore: An Encyclopedia*, ed. Margaret A. Mills, P. J. Claus and S. Diamond (New York, 2003).

55 Ibid.

56 www.lifeofguangzhou.com, accessed 28 September 2011.

5 THE MEDICINAL GOAT

1 Victoria Sherrow, *Encyclopedia of Hair: A Cultural History* (Westport, CT, 2006), p. 174.

2 Pliny the Elder, *Natural History*, vol. XXX, ch. 10.

3 Ibid., vol. XXVIII, ch. 48.

4 Ibid., ch. 52.

5 Ibid., ch. 42.

6 Ibid.

7 E. Pettit, *Anglo-Saxon Remedies, Charms and Prayers from British Library MS Harley 585. The Lacnunga* (2001), vol. I.

8 W. D. Hand, *Magical Medicine* (Berkeley, CA, 1980), p. 298.

9 Ibid.

10 Spencer L. Rogers, 'Madstones', *Ethnic Technology Notes*, VI, San Diego Museum of Man (San Diego, CA, 1971).

11 C. Elgood, 'A Treatise on the Bezoar Stone by the Late Mahmud Bin Masud the Imad-ul-din the Physician of Ispahan', *Annals of Medical History*, 7 (1935), pp. 73–80.

12 Ibid.

13 Stephen Paget, *Ambroise Paré and his Times, 1510–1590* (New York, 1897), pp. 186–7.

14 Frederick Slare, *Experiments and Observations upon Oriental and Other Bezoar Stones, Which Prove Them To Be No Use in Physick* (London, 1715).

15 Rogers, 'Madstones'.

16 J. K. Rowling, *Harry Potter and the Half-blood Prince* (London, 2005).

17 Robert Zingg, 'Alpine Ibex (*Capra ibex ibex*)', in *Grzimek's Encyclopedia of Mammals Volume 5*, ed. S. P. Parker (Columbus, OH, 1990), pp. 514–23.

18 G. Kenneth Whitehead, *The Wild Goats of Great Britain and Ireland* (Newton Abbot, 1972), p. 66.

19 F. Hamel, *Human Animals* (London, 1915).

20 Henry Stephen Holmes-Pegler, *The Book of the Goat; Containing Full Particulars of the Various Breeds of Goats and their Profitable Management*, 4th edn (London, 1909), pp. 254–5.

21 Tsering Thakchoe Drungtso, *Tibetan Medicine* (Dharamasala, 2004).

22 S. R. Sudarshan, *Encyclopaedia of Indian Medicine: Diseases and their Cures*, vol. VI (Bangalore, 2005). Many references throughout book.

23 P. Padmanabhan and K. A. Sujana, 'Animal Products in Traditional Medicine From Attappady Hills of Western Ghats', *Indian Journal of Traditional Knowledge*, VII (2008), pp. 326–9.

24 P. Oudhia, 'Traditional Medicinal Knowledge About Excreta of Different Animals Used to Treat Many Common Diseases in Chhattisgarh, India', www.botanical.com, 2003.

25 Jharna Chakravorty, V. Benno Meyer-Rochow and Sampat Ghosh, 'Vertebrates Used for Medicinal Purposes by Members of the Nyishi and Galo Tribes in Arunachal Pradesh (North-East India)', *Journal of Ethnobiology and Ethnomedicine*, VII (2011), pp. 13–27.

26 R.R.N. Alves and I. L. Rosab, 'Zootherapy Goes to Town: The Use of Animal-based Remedies in Urban Areas of NE and N Brazil', *Journal of Ethnopharmacology*, CXIII (2007), pp. 541–55; R.R.N. Alves and H. N. Alves, 'The Faunal Drugstore: Animal-based Remedies Used in Traditional Medicines in Latin America', *Journal of Ethnobiology and Ethnomedicine*, VII (2011), p. 9.

27 E. Lev and Z. Amar, 'Ethnopharmacological Survey of Traditional Drugs Sold in the Kingdom of Jordan', *Journal of Ethnopharmacology*, LXXXII (2002), pp. 131–45.

28 Clinton Bailey, 'Bedouin Religious Practices in Sinai and the
   Negev', *Anthropos*, 77 (1982), pp. 65–88.
29 Hand, *Magical Medicine*, p. 298.
30 D. P. Jaroli, Madan Mohan Mahawar and Nitin Vyas, 'An
   Ethnozoological Study in the Adjoining Areas of Mount Abu
   Wildlife Sanctuary, India', *Journal of Ethnobiology and
   Ethnomedicine*, VI (2010), p. 6.
31 Thomas Bewick, *A General History of Quadrupeds* (London, 1792),
   p. 68.
32 Samuel X. Radbill, 'The Role of Animals in Infant Feeding',
   in *American Folk Medicine*, ed. W. D. Hand (1976), pp. 21–30.
33 Ibid.
34 J. B. Tracey, *Infant Feeding: The Simple Method*, British Goat Society
   Leaflet no. 43; J. B. Tracey, 'Goats' Milk in Infant Feeding', in
   The Goatkeeper's Guide, ed. Jill Salmon (Newton Abbot, 1976),
   appendix C.
35 Reviewed in D. S. Ziegler, et al., 'Goats' Milk Quackery', *Journal
   of Paediatric Child Health*, XLI (2005), pp. 569–71.
36 R. J. Blackham, 'Goat's Milk for Infants', *British Medical Journal*,
   II/MMCCCLXXXII (1906), pp. 452–3.
37 S. Basnet, et al., 'Fresh Goat's Milk for Infants: Myths and
   Realities – A Review', *Pediatrics*, CXXV (2010), pp. e973–7; L. S.
   Taitz and T. L. Armitage, 'Goat's Milk for Infants and Children',
   British Medical Journal, CCLXXXVIII (1984), pp. 428–9.
38 Ziegler et al., 'Goats' Milk Quackery', pp. 569–71.

6 THE TWENTY-FIRST-CENTURY GOAT

1 'Nigeria Police Hold "Robber" Goat', *BBC News*,
  http://news.bbc.co.uk, accessed 23 January 2009.
2 Declan Walsh, 'Pakistan President Asif Ali Zardari "Practises
  Animal Sacrifice"' *Guardian*, 28 January 2010.
3 David Wilkes, 'Billy the Goat Retires as Welsh Regiment Mascot',
  Daily Mail, 20 May 2009.
4 At www.nationalarchives.gov.uk.

5 G. Kenneth Whitehead, *The Wild Goats of Great Britain and Ireland* (Newton Abbot, 1972), p. 98.

6 See www.chicago-cubs-fan.com/Chicago-Cubs-Curse.html

7 The RAF unofficial gossip site is at www.e-goat.co.uk.

8 Hennes's webcam can be viewed at axl22.de/webcam/webcamfc.html.

9 The U.S. Naval Academy website has the full history of its goat mascots, www.usna.edu.

10 More details of this rather unpleasant spectacle are available at www.afghanistan.org.

11 See Marshall County Goat Festival website at www.goatsmusicandmore.com.

12 See Spindale Dairy Goat Festival website at www.goatfestival.com.

13 See Falmouth Goat Races website at www.falmouthgoatrace.org.

14 Skyros Island website has pictures of the festival at www.skyros.com and you can see the goat dance on YouTube by searching for 'apokreis, Skyros'.

15 For more on the *julebukk*, see www.mylittlenorway.com.

16 The Gävle Goat official website is at www.christmasgoat. www.visitgavle.se/sv/the-gavle-goat.

17 The Tobago Goat Race website is at www.visittobago.gov.tt.

18 The Spitalfields City Farm Goat Race website is at www.thegoatrace.org.

19 The Ugandan Goat Race website is at www.thegoatraces.com.

20 FARM-Africa, the charity that started the charity goat idea, can be found online at www.farmafrica.org.uk.

21 You can buy a charity goat at www.farmafricapresents.org.uk.

Select Bibliography

Baskin, L. and K. Danell, *Ecology of Ungulates: A Handbook of Species in Eastern Europe and Northern and Central Asia* (Heidelberg, 2003)

Holmes-Pegler, Henry Stephen, *The Book of the Goat; Containing Full Particulars of the Various Breeds of Goats and their Profitable Management*, 4th edn (London, 1909)

Porter, Valerie, *Goats of the World* (Ipswich, 1996)

Shackleton, D. M., *Wild Sheep and Goats and their Relatives: Status Survey and Conservation Action Plan for Caprinae. IUCN Report* (1997)

Solaiman, Sandra G., *Goat Science and Production* (Ames, IA, 2010)

Weaver, Sue, *The Backyard Goat* (North Adams, MA, 2011)

Whitehead, G. Kenneth, *The Wild Goats of Great Britain and Ireland* (Newton Abbot, 1972)

Associations and Websites

BIOLOGY AND EVOLUTION
www.ultimateungulate.com

GOAT SOCIETIES
British Goat Society: www.allgoats.com
U.S. Goat Society: www.americangoatsociety.com

GOAT-KEEPING
For information about keeping goats:
www.goats.co.uk

GOLDEN GUERNSEY GOATS AND SOCIETY
www.goldenguernseygoat.org

RANDOM FACTS
Some very silly pictures and random facts about goats:
www.swampyacresfarm.com/RandomGoatFacts

Acknowledgements

Writing *Goat* has been an instructive and most enjoyable experience. Michael Leaman and Jonathan Burt provided sound advice and constructive criticism, for which I am very grateful. Their comments have improved the final product immeasurably.

I am most grateful to the friends and colleagues at QMUL who generously supplied me with information and anecdotes, and who read and commented on the manuscript: Simon Joel, Tilly Tansey, Virginia Davis, Nick Croft, Carol Rennie, Chris Poutain, Rudiger Goerner and Jeremy Hicks. Many thanks to Christopher Dyer, for helping me to understand the medieval farming system, and to David Bartle of the Haberdashers' Company for a discussion of the coats of arms. Thank you also to Rustom Battiwalla, whose filthy Pashtun saying took a lot of tracking down, and which never fails to raise a laugh.

I am grateful also to Nick Wright and his team at Barts and The London School of Medicine and Dentistry, which gave me the time to write at a time when I most needed it.

A specially big thank you to my family, Peter, Ben and Michaela, who provided constant support, encouragement, tea and hugs.

Photo Acknowledgements

The author and the publishers wish to express their thanks to the below sources of illustrative material and /or permission to reproduce it.

Alamy: p. 16 (Pegaz); Marco Angelli: p. 142; Nino Barbieri: p. 24; Picture by Ardo Beltz: p. 11 bottom right; Bigstock: p. 6; Bodleian Library, Oxford: pp. 52, 122; © The Trustees of the British Museum: pp. 13 centre right, 50 top left, 82, 93, 125; Corbis: p. 136 (S.SABAWOON/epa); Clio20: p. 123; Darklich14: p. 39; Delta-9: p. 14 top left; kind permission of Paola Demattè: p. 44 top left; images reproduced with kind permission of Farm Africa: pp. 43, 141 top right; Fitzwilliam Museum, Oxford: pp. 15, 34, 46, 57, 114, 128; Cgoodwin: p. 140; Guano: p. 135; kind permission of the Haberdashers' Company p. 14 centre left; Brent Hoffman: pp. 20, 23, 32, 81; © Ladybird Books Ltd, 1968, illustration used under licence from Ladybird Books Ltd: p. 115; Library of Congress, Washington, DC: pp. 9, 18, 66, 91; Ian and Frances Mason: p. 75; Medavia: p. 27; © 2014 image copyright The Metropolitan Museum of Art / Art Resource / Scala, Florence: p. 44 top centre; Incnis Mrsi: p. 129; © 2014 Digital Image, The Museum of Modern Art, New York / Scala, Florence: p. 13 bottom left (© Succession Picasso / DACS, London 2014); National Archives and Records Administration, Washington, DC: p. 63; The National Gallery, London: pp. 96, 99, 108; courtesy of the Natural History Society of Northumbria: p. 56; Nature Picture Library: pp. 28 (Igor Shpilenok), 85 (Dave Watts); Marie-Lan Nguyen (2011): p. 97; Tony Nordin: p. 136; Numismatic Guaranty Corporation: p. 127; Rare Breeds Survival Trust, Warwickshire: p. 61; Redtigeryx: p. 74; Rex Features: p. 10 (Roger-Viollet);

Index